Life in the Time of a Pandemic: Social, Economic, Health and Environmental Impacts of COVID-19—Systems Approach Study

Life in the Time of a Pandemic: Social, Economic, Health and Environmental Impacts of COVID-19—Systems Approach Study

Editors

Oz Sahin
Russell Richards

MDPI • Basel • Beijing • Wuhan • Barcelona • Belgrade • Manchester • Tokyo • Cluj • Tianjin

Editors
Oz Sahin
Griffith University
Australia

Russell Richards
University of Queensland
Australia

Editorial Office
MDPI
St. Alban-Anlage 66
4052 Basel, Switzerland

This is a reprint of articles from the Special Issue published online in the open access journal *Systems* (ISSN 2079-8954) (available at: https://www.mdpi.com/journal/systems/special_issues/systems_COVID-19).

For citation purposes, cite each article independently as indicated on the article page online and as indicated below:

LastName, A.A.; LastName, B.B.; LastName, C.C. Article Title. *Journal Name* **Year**, *Volume Number*, Page Range.

ISBN 978-3-0365-3935-5 (Hbk)
ISBN 978-3-0365-3936-2 (PDF)

Contents

About the Editors

Oz Sahin is systems modeller with specialised expertise in a range of modelling approaches to problems of natural resource management, risk assessment, and climate change adaptation. His research interests include sustainable built environment; climate change adaptation risk assessment; ecosystem based approaches; integrated water, energy, and climate modelling; integrated decision support systems using coupled system dynamics and GIS modelling; Bayesian Network modelling; multiple criteria decision analysis; and operational research methods and models. He is a senior researcher at the Griffith Climate Change Response Program and Cities Research Institute. In addition to his research, he also coordinates/teaches online courses with the School of Environment and Sciences at Griffith University.

Russell Richards is a systems modeller with experience in developing and applying models to socioecological systems in Australia and internationally. His research portfolio includes working with decision makers and stakeholders from a range of industries and disciplines, including government, public health, tourism, and fisheries. He has expertise in developing models that draw upon and combine systems thinking, system dynamics modelling, agent-based modelling, and statistical techniques. Russell is a lecturer at the University of Queensland Business School.

Editorial

Introduction to the Special Issue "Life in the Time of a Pandemic: Social, Economic, Health and Environmental Impacts of COVID-19—Systems Approach Study"

Oz Sahin [1,2,3,*] and Russell Richards [2,4,*]

1 School of Engineering and Built Environment, Griffith University, Southport, QLD 4222, Australia
2 Cities Research Institute, Griffith University, Southport, QLD 4222, Australia
3 Griffith Climate Change Response Program, Griffith University, Southport, QLD 4222, Australia
4 UQ Business School, University of Queensland, Brisbane, QLD 4072, Australia
* Correspondence: o.sahin@griffith.edu.au (O.S.); r.richards@business.uq.edu.au (R.R.)

Citation: Sahin, O.; Richards, R. Introduction to the Special Issue "Life in the Time of a Pandemic: Social, Economic, Health and Environmental Impacts of COVID-19—Systems Approach Study". *Systems* **2022**, *10*, 36. https://doi.org/10.3390/systems10020036

Received: 9 March 2022
Accepted: 10 March 2022
Published: 15 March 2022

Publisher's Note: MDPI stays neutral with regard to jurisdictional claims in published maps and institutional affiliations.

The preambles in many of the articles in this Special Issue have highlighted how COVID-19 has affected, and is continuing to affect, the way that individuals, groups, organisations and countries operate. The health implications of COVID-19 have seen decision makers take drastic interventions to address the threat to health associated with this disease. However, this has had cascading effects on other aspects of society and the environment. As expressed in the information provided for this Special Issue, "Life in the Time of a Pandemic: Social, Economic, Health and Environmental Impacts of COVID-19—Systems Approach Study", the role of governments around the world has been aimed at containing and reducing the socioeconomic impacts of COVID-19; however, their respective responses have not been consistent. Some 18 months after our call for papers, COVID-19 continues to challenge how governments and individuals manage this pandemic.

The resulting Special Issue from our call comprises nine research papers. These nine papers reflect a good diversity of foci and methodologies, ranging from conceptual/qualitative papers that provide exploration of networks to data-driven models that take advantage of the proliferation of data that have been created during the pandemic, through to fully parameterised deterministic density-based and agent-based process modelling.

In this Special Issue, the first article, by Sahin et al. [1], provides the broad context for the complexity of the COVID-19 pandemic, highlighting the multifaceted, and intrinsically intertwined characteristics of this 'system'. This communication paper produced a preliminary causal loop diagram (CLD) that endeavoured to map out this wicked complexity and advocated the need for considering 'deep leverage' (interventions) points as part of the management plans. CLDs are a commonly used technique in systems thinking, providing an illustrative map of network causality for a system. The second article, by Strelkovskii and Rovenskaya [2], thus provides a timely critique of CLDs that have been developed for COVID-19, including that developed by Sahin et al. [1], producing a set of good practices for creating and presenting these causal maps.

Unsurprisingly, disease models using the ubiquitous susceptible–infected–recovered (SIR) or susceptible–exposed–infected–recovered (SEIR) frameworks are featured in two papers (Bärwolff [3]; Brereton and Pedercini [4]). Such density-based dynamic models enable the evolving nature of 'what if' health-management scenarios to be tested over a period of time from the safety of a numerical playground. Specifically, it has been used in these two papers to assess the effectiveness of 'lockdowns' against indicators such as infection rates, as explored by these two papers.

Several papers drew upon the large number of data that have been produced throughout the COVID-19 pandemic to undertake data-driven analysis. For example, Bertone et al. [5] parameterised Naïve Bayesian networks with such data in their analysis of the impact of lockdown timing on case and mortality numbers. Whilst these data have proven to be a goldmine when it comes to creating data-driven (and process-based) models, as highlighted in many of the articles in this issue, it has also created a 'proliferation of multiple views' as investigated by Stella. In this paper, Stella [6] used an analysis of media responding to the WHO declaration of the global pandemic and semantic frame theory with emotional profiling to reconstruct the 'plurality of views and emotions' elicited from this declaration. This showed that this declaration elicited a wide spectrum of perceptions, including anger and grief, but also trust.

From a business perspective, COVID-19 has severely restricted mobilisation, which has disrupted traditional businesses operations. However, it is recognised that this has also created opportunities within the digital landscape. For example, the article by Sorooshian [7] focused on 'change readiness' for the digitisation of tourism. A key finding was that business tourism and event tourism were the most ready for this to occur. The article by Sindhu and Mor [8] highlighted how COVID-19 had facilitated an increase dependence of consumers using digital platforms and identified the importance of measurement and evaluation strategies, and customer as co-creators, as enabling factors for branded content.

The final paper (Harré et al. [9]) presented a comprehensive use of agent-based modelling to evaluate a variety of different mechanisms through which crises can propagate from the micro-economic behaviour through to an economy's aggregate dynamics. This includes an exploration of the impacts of the government's COVID-19 policy on Australia's housing market.

Due to the timing of this Special Issue, much of the focus of these nine papers has been on the dynamics of COVID-19 during 2020 and early–mid 2021. As we enter 2022, vaccination programs are well established in many countries (particularly the 'Global North') and the narrative of 'opening up' and 'living with COVID' is becoming an increasing catchcry. However, COVID-19 is still globally pervasive with reported infection rates higher now than they were during 2020–2021, and decision makers continue to grapple with balancing health and economics.

Conflicts of Interest: The authors declare no conflict of interest.

References

1. Sahin, O.; Salim, H.; Suprun, E.; Richards, R.; MacAskill, S.; Heilgeist, S.; Rutherford, S.; Stewart, R.A.; Beal, C.D. Developing a Preliminary Causal Loop Diagram for Understanding the Wicked Complexity of the COVID-19 Pandemic. *Systems* **2020**, *8*, 20. [CrossRef]
2. Strelkovskii, N.; Rovenskaya, E. Causal Loop Diagramming of Socioeconomic Impacts of COVID-19: State-of-the-Art, Gaps and Good Practices. *Systems* **2021**, *9*, 65. [CrossRef]
3. Bärwolff, G. Mathematical Modeling and Simulation of the COVID-19 Pandemic. *Systems* **2020**, *8*, 24. [CrossRef]
4. Brereton, C.; Pedercini, M. COVID-19 Case Rates in the UK: Modelling Uncertainties as Lockdown Lifts. *Systems* **2021**, *9*, 60. [CrossRef]
5. Bertone, E.; Luna Juncal, M.J.; Prado Umeno, R.K.; Peixoto, D.A.; Nguyen, K.; Sahin, O. Effectiveness of the Early Response to COVID-19: Data Analysis and Modelling. *Systems* **2020**, *8*, 21. [CrossRef]
6. Stella, M. Cognitive Network Science Reconstructs How Experts, News Outlets and Social Media Perceived the COVID-19 Pandemic. *Systems* **2020**, *8*, 38. [CrossRef]
7. Sorooshian, S. Implementation of an Expanded Decision-Making Technique to Comment on Sweden Readiness for Digital Tourism. *Systems* **2021**, *9*, 50. [CrossRef]
8. Sindhu, S.; Mor, R.S. Modelling the Enablers for Branded Content as a Strategic Marketing Tool in the COVID-19 Era. *Systems* **2021**, *9*, 64. [CrossRef]
9. Harré, M.S.; Eremenko, A.; Glavatskiy, K.; Hopmere, M.; Pinheiro, L.; Watson, S.; Crawford, L. Complexity Economics in a Time of Crisis: Heterogeneous Agents, Interconnections, and Contagion. *Systems* **2021**, *9*, 73. [CrossRef]

Communication

Developing a Preliminary Causal Loop Diagram for Understanding the Wicked Complexity of the COVID-19 Pandemic

Oz Sahin [1,2,3,*], Hengky Salim [1,2], Emiliya Suprun [1,2], Russell Richards [2,4], Stefen MacAskill [1,2], Simone Heilgeist [1,2], Shannon Rutherford [5], Rodney A. Stewart [1,2] and Cara D. Beal [1,2,5]

[1] School of Engineering and Built Environment, Griffith University, Brisbane, QLD 4222, Australia; hengky.salim@griffithuni.edu.au (H.S.); e.suprun@griffith.edu.au (E.S.); stefen.macaskill@griffithuni.edu.au (S.M.); simone.heilgeist@griffithuni.edu.au (S.H.); r.stewart@griffith.edu.au (R.A.S.); c.beal@griffith.edu.au (C.D.B.)
[2] Cities Research Institute, Griffith University, Brisbane, QLD 4222, Australia; r.richards@business.uq.edu.au
[3] Griffith Climate Change Response Program, Griffith University, Brisbane, QLD 4222, Australia
[4] UQ Business School, University of Queensland, Brisbane, QLD 4072, Australia
[5] School of Medicine, Gold Coast campus, Griffith University, Brisbane, QLD 4222, Australia; s.rutherford@griffith.edu.au
* Correspondence: o.sahin@griffith.edu.au

Received: 9 May 2020; Accepted: 17 June 2020; Published: 18 June 2020

Abstract: COVID-19 is a wicked problem for policy makers internationally as the complexity of the pandemic transcends health, environment, social and economic boundaries. Many countries are focusing on two key responses, namely virus containment and financial measures, but fail to recognise other aspects. The systems approach, however, enables policy makers to design the most effective strategies and reduce the unintended consequences. To achieve fundamental change, it is imperative to firstly identify the "right" interventions (leverage points) and implement additional measures to reduce negative consequences. To do so, a preliminary causal loop diagram of the COVID-19 pandemic was designed to explore its influence on socio-economic systems. In order to transcend the "wait and see" approach, and create an adaptive and resilient system, governments need to consider "deep" leverage points that can be realistically maintained over the long-term and cause a fundamental change, rather than focusing on "shallow" leverage points that are relatively easy to implement but do not result in significant systemic change.

Keywords: COVID-19; pandemic; wicked problem; systems approach; leverage points

---

1. Introduction

The COVID-19 pandemic has emerged as a problem of wicked complexity for policy makers internationally [1]. The virus and its necessary management strategies have thrown many countries into economic recession [2] and exacerbated existing social problems such as health care access, unemployment and inequality. A few countries have rapidly responded to the pandemic and have had success in its early containment, yet many countries have scrambled to implement interventions and measures when major implications of the disease started to appear.

Policy makers around the world have been mainly focusing on two key responses, namely, virus containment and financial measures for cushioning the resulting economic impact (i.e., jobs subsidies, unemployment benefits, government supported loans, etc.) [3]. However, they have been conducting these assessments separately from a "wait-and-see" perspective and often have not systematically examined this problem with consideration to a feasible and sustainable long-term strategy for managing the pandemic. Many countries have failed to learn from past epidemics of coronaviruses such as Severe

Acute Respiratory Syndrome (SARS) and Middle East Respiratory Syndrome (MERS), where these types of viruses have historically long incubation periods [4]. A few countries have responded to this pandemic rapidly [5] with the readiness to sacrifice an early economic loss to prevent worse long-term economic impacts that would occur if the virus had spiralled out of control.

Systems thinking is a framework that can help policy makers to better understand the big picture through identifying the multi-faceted consequences of decision making in order to better weigh options and design the most effective strategies to manage the impacts of unintended consequences [6–8]. Effectively containing the virus and keeping the mortality rate low while maintaining economic, social and environmental goals is of importance in effectively managing this pandemic. The aim of this communication piece is to visualise the complexity in managing the COVID-19 pandemic through a systems lens by identifying the interconnectivity between health, economic, social and environmental aspects. This was explored via the development of a preliminary causal loop diagram (CLD) to identify important feedback loops. In the systems thinking field, CLD is a powerful tool for dealing with complex problems which has the ability to uncover the underlying feedback structures and leverage points in a system [9–11]. Moreover, causal loop modelling has been widely applied in health systems research [12–14].

2. A Wicked Complexity

In an increasingly connected world, the actions of individuals and governments and their resulting consequences are deeply entwined within the socio-economic and environmental systems. Recognising that the impacts of the COVID-19 pandemic transcend many boundaries (e.g., health, communities, science, politics, environment and economics) will help policy makers to determine the "right" intervention in a timely manner and implement additional interventions to reduce negative consequences. A CLD was developed to represent this complex problem through the identification of cause-and-effect links and feedback loops (Figure 1). This preliminary CLD was a product of the collective knowledge of the authors supported by geographical data by Johns Hopkins University [15] and a review of various governments' responses to the COVID-19 pandemic. The process of developing the preliminary CLD is presented in Appendix A.

Interventions (i.e., leverage points) are central to mitigating this pandemic. International travel restrictions, business restrictions, effectiveness of health crisis management, testing, awareness and social distancing campaigns and economic stimulus packages are among the interventions that have been implemented in many countries. Each intervention, undoubtedly, will have a trade-off between aspects of the system. An example is where the mandated "social distancing" rules can have a significant and immediate impact on business operations with potentially long-term economic and social consequences. Furthermore, delays will also exist in the system to indicate the time required for an intervention to be implemented or for a change to have an impact on the overall system. There is also a delay between the infection and when the symptoms appear (i.e., incubation period) which has caused more challenges in preventing an outbreak [16].

The existence of feedback loops within the system indicates two-way relationships between actions and consequences. These feedback loops can be used to identify if an intervention is able to create a system-wide change or if there is a need to improve or introduce a new solution. There are two types of feedback loops: reinforcing and balancing loops [8,17]. Reinforcing loops are responsible for the creation of an exponential growth or decay in the system, whereas balancing loops will balance a system until an equilibrium has been achieved. The dominance of reinforcing loops in this system indicates that there are more sources of growth, erosion and failure which decision makers need to address and minimise. Many countries have failed to realise and address these reinforcing loops [18]; thus, causing a near-collapse effect in the system that is exhibited by a massive outbreak in many countries.

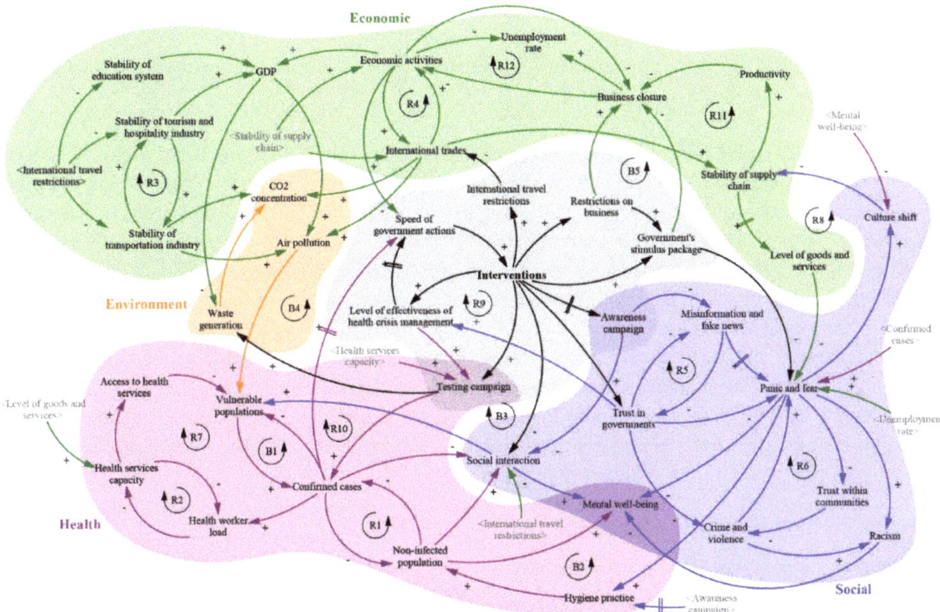

Figure 1. A preliminary causal loop diagram demonstrating the complexity of the COVID-19 pandemic environmental–health–socio–economic system.

We have seen in a very short time, how the gradual increase of interventions has led to unprecedented economic consequences [19]. The social distancing rules have created a restriction on some business operations; thus, some business closures are inevitable, consequently leading to an increase in unemployment rates (loop R12). The international and interstate travel restrictions have caused stock market volatility and have been prohibitive to international trading and mobility (loop R4 and R11). They also caused disruptions to almost every industry sector, including education systems and the interconnectivity between transport industries and the tourism and hospitality sector (loop R3). The International Monetary Fund (IMF) suggested that the world's economy will shrink by three percent this year, which is far worse than the 2008–2009 financial crisis [20]. This restrictive economic activity has led to some governments providing financial packages to affected businesses and employees (loop B5).

From the health perspective, a higher proportion of vulnerable populations will lead to a higher number of confirmed cases (loop B1). Population vulnerability is influenced by the accessibility of health services (loop R7); whilst a higher number of confirmed cases will increase health services' load (loop R2). However, fewer confirmed cases do not always reflect the actual infection rate as it is also dependent upon the effectiveness of the testing campaign (loop R10). Health care systems in many countries, particularly in developing countries, are overwhelmed with the exponential growth of cases [21–23]. Loop R9 (Figure 1) also demonstrates how a higher number of confirmed cases reinforces the speed of government actions that will lead to the introduction of additional measures. The non-infected loop can act as a reinforcing loop (R1) to reflect the number of recovered populations as well as a balancing loop (B6) if there is an increased risk of transmissions due to the increased extent of social interactions.

Conversely, positive environmental benefits should also be considered in the policy analysis system, such as the measurable decline in regional air pollution and greenhouse gas emissions due to reductions in ground and air travel. This improvement of air quality has been observed at a global scale as a result of decline in significant travel, business and social activities [24]. Although there has been a reduction on the waste generation volume due to a downward shift in economic activities, waste from personal protective equipment and testing kits will inevitably rise as the number of vulnerable in a population that need to be tested increases [25]. Furthermore, a culture shift related to panic buying will also contribute to food waste generation.

The human social network is the most challenging to manage in this pandemic with a high risk of catastrophic social order demise if inconsiderate policy is enacted [26]. This is a delicate balancing act; for example, maintaining social distancing will substantially reduce virus transmission (loop B3), however, long periods of isolation may have long lasting effects on mental well-being. Furthermore, this pandemic has caused global panic, heightened fear and eroded trust in governments and within communities. The prevalence of reinforcing loops R5 (trust), R6 (sense of security) and R8 (panic buying) reflects the increasing social issues that need to be addressed. We have seen multiple instances of societal behavioural changes such as panic buying (e.g., toilet paper and sanitiser), the emergence of organised crime, domestic violence, increased and targeted global xenophobia against certain ethnic groups [18,27] and more recently abuse of health care workers as people's fears lead to irrationality and anxiety. These social problems stem from bounded rationality and responses to their panic and fear [28] where misinformation and confirmation bias may be contributing factors. Beside its social impacts, panic and fear may also increase hygiene practices among communities (loop B2).

3. Placing Interventions in the Right Place at the Right Time

Countries have taken different roads in addressing this global pandemic, leading to an activation of different leverage points. Two questions arise "Have the most widely used global interventions so far targeted relatively ineffective leverage points? Has current intervention been focused too heavily on 'shallow' leverage points?". As Meadows stated [8], there are twelve places in enacting leverage points ranging from "shallow" to "deep" (Figure 2). "Shallow" leverage points refer to interventions that are relatively easy to implement, yet bring a non-significant systemic change, while "deep" leverage points will cause a fundamental change.

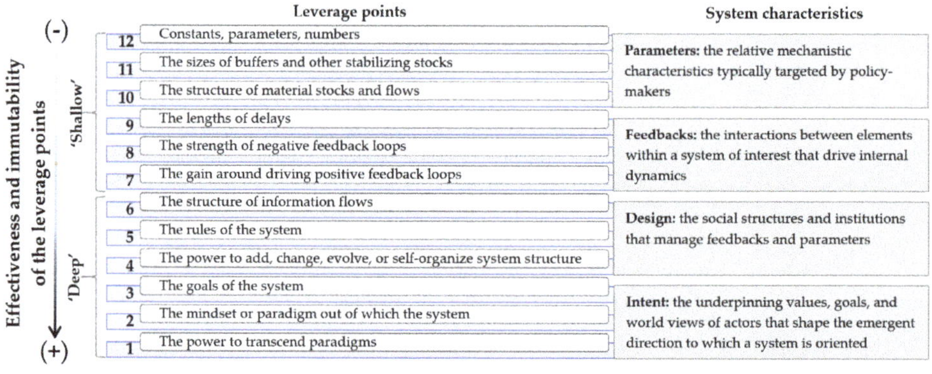

Figure 2. Leverage points to intervene a system. Adapted from Meadows [8].

Policies enacted in the interests of public health have had economic side effects. The extent, of course, will largely depend on the depth of the early intervention, the ability of governments to enact policy to limit the damage to the economy and the strength of the economy prior to the event. The travel restriction is a good example of both shallow and deep leveraging points as travel restrictions can change the direction to which a system is oriented by preventing infection being transmitted into a population by non-symptomatic carriers. Australia followed a series of largely ineffective (or shallow) international travel restrictions in the early stages where a ban was in place for flights from the source of the outbreak. It was not until it enacted a complete ban for all international and interstate flights (i.e., deeper leverage point) that the spread of new cases appeared to slow down.

While encouraging social distancing and vigilance in personal hygiene are critical measures to reduce the risk of human-to-human transmission, it is challenging for the general public to consciously change their behaviour overnight and maintain these behaviours long-term [29]. Regardless of laws put in place, there remains a threat of public complacency, particularly once infection rates decrease. Government policies need to consider deeper leverage points that can be realistically maintained over the long-term, as infection rates trend up and down. Public confidence and trust in governance may be negatively impacted with regular "shallow", knee-jerk daily or weekly rule changes.

4. Lessons Learned

A systemic change could transcend the "wait and see" paradigm into a more proactive approach that is imperative for creating an adaptive and resilient system. Regardless of the approach taken, countries which have demonstrated a capacity to assess this problem systemically and comprehensively over various time horizons will emerge from this crisis in a much better position than those that have just tackled each incremental problem in an isolated and knee-jerk manner. It is possible that with a more proactive approach in implementing "deeper" interventions (deep leverage point), governments can be "flattening the curve" more effectively; consequently, limiting the impact of economic recession and associated socio-economic difficulties.

Government interventions will always struggle to completely prevent all types of virus transmission over the long term. However, effective government decisions must consider strategies that reduce infection rates, while dynamically accounting for the economic, social and environmental goals. Putting our "systems thinking" hat on to tackle this wicked problem, will help us to understand that there are always ever-moving and conflicting goals existing in any system, and will help those policy architects to develop best-practice (and deeper) interventions that will help to minimise unintended negative outcomes.

This communication piece reports on the development of a preliminary CLD which depicts the complexity and the multi-faceted nature of the COVID-19 pandemic from health, economic, social and environmental perspectives. This piece aims to demonstrate the COVID-19 pandemic complexity across health–socio–economic–environmental boundaries using a systems thinking visual as a precursor for the special issue "Life in the Time of a Pandemic: Social, Economic, Health and Environmental Impacts of COVID-19—Systems Approach Study" in the Systems journal. It is not intended to provide a full explanation of this issue but rather provide an example of how visualising the complexity of a system can help us to identify leverage points and the key important trade-offs that exist in the system. There remains a need to develop a system dynamics (SD) model that will be able to quantify this system. Such a model will assist policy makers in enacting an effective strategy for preparing nations in defending themselves against future pandemics by revealing the complexity, dynamic behaviour and trade-offs between different objectives. We would be foolish to not learn from the lessons around complexity and system interactions that this COVID pandemic has presented.

Author Contributions: O.S., H.S., E.S., R.R., S.M., and S.H. conducted stakeholder workshops, developed the preliminary causal loop diagram and drafted the manuscript. S.R., R.A.S., and C.D.B. conducted stakeholder workshops and improved the manuscript. All authors have read and agreed to the published version of the manuscript.

Funding: This research received no external funding.

Conflicts of Interest: The authors declare no conflict of interest.

Glossary

Variable Name	Description
Access to health services	Access into the health care system
Air pollution	The amount of harmful substances in earth's atmosphere
Awareness campaign	Marketing effort to educate individuals about an issue (e.g., need in regular hand washing, coverage of coughing/sneezing, usage of tissues and bin tissues, etc.)
Business closure	Closure of businesses due both temporarily and due to bankruptcy
CO_2 concentration	Concentration of carbon dioxide in earth's atmosphere
Confirmed cases	Positive tested population
Crime and violence	Intentional harm
Culture shift	Shift in communities' culture (e.g., panic buying, business culture, etc.)
Economic activities	Stable manufacture of goods and the provision of services
GDP	Gross domestic product
Government's stimulus package	Government's financial assistance to support businesses, households, and individuals
Health care worker load	Number of patients per health care professional
Health services capacity	Availability of facial masks, hospital beds, medication, treating medical staff, public health services
Hygiene practice	Regular hand washing, coverage of coughing/sneezing, usage of tissues and bin tissues
International trades	Export and import of goods and services
International travel restrictions	Travel ban to international flights to curb imported cases
Interventions	Action plan of the government to controlling pandemic and its impacts
Level of effectiveness of government health crisis actions	Effective operational action plan of the government; innovative steps to enable an effective intervention
Level of goods and services	Goods and services available in the market
Mental well-being	Social and emotional well-being of individuals
Misinformation and fake news	Pseudo-news, deliberate disinformation, conspiracy theories or hoaxes spread via traditional news media or online social media
Non-infected population	Fatalities, recovered, negative tested, non-tested population
Panic and fear	Sudden anxiety, hysterical and irrational behaviour
Productivity	Rate of goods and services being produced
Racism	Prejudice, discrimination, or hatred directed at someone because of their colour, ethnicity or national origin
Restrictions on business	Temporarily closure of non-essential businesses, 'take-away' only policy
Social interaction	Ability to meet (an)other individual(s)
Speed of government actions	Coordinated and timely operational action plan of the government to address health crisis
Stability of education system	Uninterrupted work of education institutes, high number of enrolled international students at universities
Stability of supply chain	Uninterrupted distribution of goods and services, (i.e., no delays), availability of goods and services available in the market
Stability of tourism and hospitality industry	Uninterrupted flow of both international and domestic visitors, stable work of hospitality businesses and events
Stability of transportation industry	Stable work of airlines, train services, shipping industries
Testing campaign	Campaign to promote public awareness about COVID-19 testing if they have any symptom
Trust in governments	Community trust and confidence towards parliament, the cabinet, the civil service, local councils, political parties, politicians
Trust within communities	The degree of trust towards a certain group of people
Unemployment rate	Share of the labour force that is jobless
Vulnerable populations	Elderly, socioeconomically disadvantaged (uninsured, homeless), individuals with a pre-existing medical condition
Waste generation	The amount of waste generated by households, industries and health systems

Appendix A

Table A1. Development of the preliminary CLD through expert workshops.

Modelling Workshops	Purpose	Date and Meeting Format	Number of Participants	Participants' Area of Expertise **	Confirmed COVID-19 Cases, Australia
Workshop 1: Problem scoping	• Identification of the key variables in regard to impacts of the COVID-19 pandemic • Confirmation of the system boundary: COVID-19 pandemic environmental-health-socio-economic system	March 10 2020; Face-to-face	5	PH, S, CC, EM, ST, MS, G, E, EngM	116
Workshop 2: Model conceptualisation	• Identification of relationships among identified variables • Construction of a preliminary conceptual model represented as a CLD	March 24 2020; Face-to-face	7	PH, S, CC, EM, ST, MS, G, EngM, E, F, B	2317
Workshop 3: Model confirmation	• Refinement of the initial CLD generated from the previous modelling workshop	March 29 2020; Video conferencing *	7	PH, S, CC, EM, ST, MS, EngM, G, E, F, B	4163
Workshop 4: Model confirmation (Cont.)	• Identification of feedback loops • Identification of interventions (i.e., leverage points)	April 5 2020; Video conferencing *	4	S, CC, EM, ST, MS, EngM, G, E	5750

Notes: * Due to COVID-19 pandemic restrictions last two workshops were held in the digital realm via video conferencing platforms. ** PH = Public health; S = Sustainability; CC = Climate change; EM = Environmental management; ST = Systems thinking; MS = Modelling and simulation; G = Governance; E = Economics; EngM = Engineering Management; F = Finance; B = Business.

References

1. Norman, C.D. Health promotion as a systems science and practice. *J. Eval. Clean. Pract.* **2009**, *15*, 868–872. [CrossRef] [PubMed]
2. Wearden, G.; Jolly, W. IMF: Global Economy Faces Worst Recession Since the Great Depression—As It Happened. *The Guardian.* 15 April 2020. Available online: https://www.theguardian.com/business/live/2020/apr/14/stock-markets-china-trade-global-recession-imf-forecasts-covid-19-business-live (accessed on 20 April 2020).
3. Anderson, R.M.; Heesterbeek, H.; Klinkenberg, D.; Hollingsworth, T.D. How will country-based mitigation measures influence the course of the COVID-19 epidemic? *Lancet* **2020**, *395*, 931–934. [CrossRef]
4. Prompetchara, E.; Ketloy, C.; Palaga, T. Immune responses in COVID-19 and potential vaccines: Lessons learned from SARS and MERS epidemic. *Asian Pac. J. Allergy* **2020**, *38*, 1–9. [CrossRef]
5. Barron, L. What We Can Learn from Singapore, Taiwan and Hong Kong about Handling Coronavirus. *Time.* 13 March 2020. Available online: https://time.com/5802293/coronavirus-covid19-singapore-hong-kong-taiwan/ (accessed on 22 April 2020).
6. Da Luke, D.A.; Stamatakis, K.A. System science methods in public health: Dynamics, networks, and agents. *Annu. Rev. Public Health* **2012**, *33*, 357–376. [CrossRef] [PubMed]
7. Suprun, E.; Sahin, O.; Stewart, R.A.; Panuwatwanich, K. Model of the Russian Federation construction innovation system: An integrated participatory systems approach. *Systems* **2016**, *4*, 29. [CrossRef]
8. Meadows, D.H. *Thinking in Systems*; Chelsea Green Publishing: White River Junction, VT, USA, 2008.
9. Bureš, V. A method for simplification of complex group causal loop diagrams based on endogenisation, encapsulation and order-oriented reduction. *Systems* **2017**, *5*, 46. [CrossRef]
10. Agnew, S.; Smith, C.; Dargusch, P. Causal loop modelling of residential solar and battery adoption dynamics: A case study of Queensland, Australia. *J. Clean. Prod.* **2018**, *172*, 2363–2373. [CrossRef]
11. Garrity, E.J. Using systems thinking to understand and enlarge mental models: Helping the transition to a sustainable world. *Systems* **2018**, *6*, 15. [CrossRef]
12. Rees, D.; Cavana, R.Y.; Cumming, J. Using cognitive and causal modelling to develop a theoretical framework for implementing innovative practices in primary healthcare management in New Zealand. *Health Syst.* **2018**, *7*, 51–65. [CrossRef]
13. Lembani, M.; de Pinho, H.; Delobelle, P.; Zarowsky, C.; Mathole, T.; Ager, A. Understanding key drivers of performance in the provision of maternal health services in eastern cape, South Africa: A systems analysis using group model building. *BMC Health Serv. Res.* **2018**, *18*, 912. [CrossRef]
14. Littlejohns, L.B.; Baum, F.; Lawless, A.; Freeman, T. The value of a causal loop diagram in exploring the complex interplay of factors that influence health promotion in a multisectoral health system in Australia. *Health Res. Policy Syst.* **2018**, *16*, 126. [CrossRef] [PubMed]
15. Dong, E.; Du, H.; Gardner, L. An interactive web-based dashboard to track COVID-19 in real time. *Lancet* **2020**, *20*, 533–534. [CrossRef]
16. Lauer, S.A.; Grantz, K.H.; Bi, Q.; Jones, F.K.; Zheng, Q.; Meredith, H.R.; Azman, A.S.; Reich, N.G.; Lessler, J. The incubation period of coronavirus disease 2019 (COVID-19) from publicly reported confirmed cases: Estimation and application. *Ann. Intern. Med.* **2020**, *172*, 577–582. [CrossRef] [PubMed]
17. Sterman, J. *Business Dynamics: Systems Thinking and Modeling for a Complex World*; McGraw-Hill: New York, NY, USA, 2000.
18. Bradley, D.T.; Mansouri, M.A.; Kee, F.; Garcia, L.M.T. A systems approach to preventing and responding to COVID-19. *EClinicalMedicine* **2020**. [CrossRef] [PubMed]
19. McKibbin, W.J.; Fernando, R. The global macroeconomic impacts of COVID-19: Seven scenarios. *CAMA Work. Pap.* **2020**, *19*. [CrossRef]
20. International Monetary Fund. *World Economic Outlook Reports*; International Monetary Fund: Washington, DC, USA, 2020.
21. Gilbert, M.; Pullano, G.; Pinotti, F.; Valdano, E.; Poletto, C.; Boëlle, P.; D'Ortenzio, E.; Yazdanpanah, Y.; Eholie, S.P.; Pharm, M.A.; et al. Preparedness and vulnerability of African countries against importations of COVID-19: A modelling study. *Lancet* **2020**, *395*, 871–877. [CrossRef]
22. Loayza, N.V.; Pennings, S. *Macroeconomic Policy in the Time of COVID-19: A Primer for Developing Countries*; World Bank: Washington, DC, USA, 2020.

23. Grasselli, G.; Pesenti, A.; Cecconi, M. Critical care utilization for the COVID-19 outbreak in Lombardy, Italy: Early experience and forecast during an emergency response. *JAMA* **2020**. [CrossRef]

24. Muhammad, S.; Long, X.; Salman, M. COVID-19 pandemic and environmental pollution: A blessing in disguise? *Sci. Total Environ.* **2020**, *728*, 138820. [CrossRef]

25. World Health Organization. Water, Sanitation, Hygiene and Waste Management for the COVID-19 Virus. Available online: https://apps.who.int/iris/bitstream/handle/10665/331305/WHO-2019-NcOV-IPC_WASH-2 020.1-eng.pdf (accessed on 7 May 2020).

26. Galea, S.; Merchant, R.M.; Lurie, N. The mental health consequences of COVID-19 and physical distancing: The need for prevention and early intervention. *JAMA Intern. Med.* **2020**. [CrossRef]

27. Habibi, R.; Burci, G.L.; de Campos, T.C.; Chirwa, D.; Cinà, M.; Dagron, S.; Eccleston-Turner, M.; Forman, L.; Gostin, L.O.; Meier, B.M.; et al. Do not violate the International Health Regulations during the COVID-19 outbreak. *Lancet* **2020**, *395*, 664–666. [CrossRef]

28. Sterman, J.D.; Dogan, G. I'm not hoarding, I'm just stocking up before the hoarders get here: Behavioral causes of phantom ordering in supply chains. *J. Oper. Manag.* **2015**, *39–40*, 6–22. [CrossRef]

29. Van Bavel, J.J.; Baicker, K.; Boggio, P.S.; Capraro, V.; Cichocka, A.; Cikara, M.; Crockett, A.J.; Crum, A.J.; Douglas, K.M.; Druckman, J.N.; et al. Using social and behavioural science to support COVID-19 pandemic response. *Nat. Hum. Behav.* **2020**, *4*, 460–471. [CrossRef] [PubMed]

Review

Causal Loop Diagramming of Socioeconomic Impacts of COVID-19: State-of-the-Art, Gaps and Good Practices

Nikita Strelkovskii [1],* and Elena Rovenskaya [1,2]

[1] Advancing Systems Analysis Program, International Institute for Applied Systems Analysis (IIASA), Schlossplatz 1, 2361 Laxenburg, Austria; rovenska@iiasa.ac.at

[2] Faculty of Computational Mathematics and Cybernetics, Lomonosov Moscow State University, GSP-1, Leninskie Gory, 119991 Moscow, Russia

* Correspondence: strelkon@iiasa.ac.at

Abstract: The complexity, multidimensionality, and persistence of the COVID-19 pandemic have prompted both researchers and policymakers to turn to transdisciplinary methods in dealing with the wickedness of the crisis. While there are increasing calls to use systems thinking to address the intricacy of COVID-19, examples of practical applications of systems thinking are still scarce. We revealed and reviewed eight studies which developed causal loop diagrams (CLDs) to assess the impact of the COVID-19 pandemic on a broader socioeconomic system. We find that major drivers across all studies are the magnitude of the infection spread and government interventions to curb the pandemic, while the most impacted variables are public perception of the pandemic and the risk of infection. The reviewed COVID-19 CLDs consistently exhibit certain complexity patterns, for example, they contain a higher number of two- and three-element feedback loops than comparable random networks. However, they fall short in representing linear complexity such as multiple causes and effects, as well as cascading impacts. We also discuss good practices for creating and presenting CLDs using the reviewed diagrams as illustration. We suggest that increasing transparency and rigor of the CLD development processes can help to overcome the lack of systems thinking applications to address the challenges of the COVID-19 crisis.

Keywords: causal loop diagram; systems thinking; COVID-19; network theory

Citation: Strelkovskii, N.; Rovenskaya, E. Causal Loop Diagramming of Socioeconomic Impacts of COVID-19: State-of-the-Art, Gaps and Good Practices. *Systems* **2021**, *9*, 65. https://doi.org/10.3390/systems9030065

Academic Editors: Oz Sahin and Russell Richards

Received: 30 July 2021
Accepted: 30 August 2021
Published: 2 September 2021

1. Introduction

Despite a significant progress on vaccination, with almost four billion vaccine doses administered, the daily number of new COVID-19 cases worldwide is still around the 500,000 mark, and the daily number of deaths is close to 10,000 as of late July 2021 [1]. Furthermore, various new mutations of the virus, an uneven distribution of vaccines across different countries, the unwillingness of large parts of the populations in some countries to receive vaccination, as well as other factors contribute to the persistence of the COVID-19 crisis as the most pressing issue globally [2].

The COVID-19 pandemic is not only a grand challenge for the public health system, but it has also affected virtually all areas of human life. The spread of the virus, as well as various mitigation and adaptation measures have had a widespread effect on economic activity, job security, social relations, mental health, and trust in others and institutions [3]. This makes the challenge of "getting back to normal life" truly multi-dimensional and calls for an interdisciplinary approach [4]. However, multiple and potentially lagged interdependencies between various components of the affected systems are difficult to oversee and comprehend by the human brain in the absence of special tools, while the lack of a holistic perspective increases the risks of unintended adverse consequences [5,6]. Systems thinking has been suggested to unravel this challenge by accounting for essential links and feedback loops between issues that both scientists and policymakers tend to consider in isolation, creating a shared understanding of the problem and identifying potential leverage points [7,8].

Some scholars responded to this call advocating the use of systems thinking in a rather general sense [9–11], while others came up with some concrete examples of the application of systems thinking, usually through employing causal loop diagrams [2,4,12–16] or system dynamic models [17–19].

Causal loop diagramming (also termed systems mapping) is a principal qualitative system thinking tool used both inside academia and for communicating with policymakers and the general public [20]. Causal loop diagrams (CLDs) constitute a schematic description of the considered system depicting its components and the (causal) relations between them. Components are connected by directed links. Each link represents an impact (causal influence) of one component on another. The impact can be positive, in which case an increase/decrease of the state of the impacting component leads to an increase/decrease of the state of the impacted component, or negative, in which case an increase/decreases of the state of the impacting component leads to the opposite change of the state of the impacted component, i.e., a decrease/increase. CLDs are useful for formalizing mental models of individuals and groups, rapid identification of the possible drivers of the considered system's dynamics, and communicating feedback and archetypal structures in the considered system [20]. CLDs can be used as a standalone qualitative modeling tool or as a step toward developing a quantitative simulation, e.g., a system dynamics model [21].

This paper aims to review the state-of-the-art studies that construct CLDs to investigate the impact of the COVID-19 pandemic on a broader human–society–environment system. This review intends to formulate methodological as well as applied insights. Methodologically, our analysis provides observations (a) on what seems to be a common practice in research involving causal loop diagramming to analyze the socioeconomic impacts of COVID-19 from the systems perspective; (b) on major gaps in the existing CLDs that deal with systems impact of COVID-19; and (c) on what seems to be a good practice in the development, presentation, and analysis of CLDs. Observations (a), (b), and (c) can be useful for future CLD developers for benchmarking their work against the state-of-the-art, for positioning and focusing their research, and for increasing the impact of their research, respectively. The applied insights of this paper include observations that can guide quantitative model development to further analyze the multi-dimensional impacts of COVID-19 and policy-relevant observations.

The paper is organized as follows. The approach to the selection of studies for the review as well as key methods for the analysis of the selected CLD set is described in Section 2. In Section 3, we present the results of the analysis of the selected studies including a summary of the selected papers and their scope (Section 3.1), analysis of commonly and rarely used concepts across the reviewed CLDs (Section 3.2), basic network statistics of the reviewed CLDs (Section 3.3), major drivers and impacted components (Section 3.4), complexity patterns (Section 3.5), and, finally, the discussion of good practices for the CLD development, presentation and analysis as used by the authors of the reviewed CLDs (Section 3.6). Section 4 provides a discussion and conclusions.

2. Methods and Scope

To identify relevant studies, first, we conducted a formal literature search in the Scopus database using the following search query:

TITLE-ABS-KEY (COVID-19 AND (("causal loop diagram*") OR ("influence diagram*") OR ("systems map*"))).

Therefore, we also accounted for terms that are sometimes used interchangeably to CLDs, i.e., systems maps and influence diagrams.

This search yielded 12 papers. Seven out of these were discarded from the further analysis, as they focused either only on the virus spread itself, i.e., being epidemiological models, e.g., [22], or on a too-narrow phenomenon, e.g., [23] focusing on the routine childhood immunization or [24] focusing on the development of branchless banking. One of the remaining five papers was a conference paper [25] that then was developed into a journal article by the same author and contained the same CLD, so we also disregarded this conference paper from our analysis and included only the journal article [2].

Then, we also reviewed the citations of the remaining four papers and, using both Scopus and the Google Scholar database added four more works containing relevant CLDs—one journal paper, two preprints, and one blog post. Ultimately, eight studies satisfying the scope of our review were selected for a detailed analysis. These eight publications are summarized in Table 1.

Table 1. Reviewed studies (sorted by date of publication, ascending).

Authors/CLD ID	Title	Date Published	Type	Reference
(Wicher, 2020)	The COVID-19 case as an example of Systems Thinking usage	15 March 2020	Blog	[26]
(Bradley et al., 2020)	A systems approach to preventing and responding to COVID-19	28 March 2020	Paper in a peer-reviewed journal	[16]
(Sahin et al., 2020)	Developing a Preliminary Causal Loop Diagram for Understanding the Wicked Complexity of the COVID-19 Pandemic	18 June 2020	Paper in a peer-reviewed journal	[12]
(Bahri, 2020)	The Nexus Impacts of the COVID-19: A Qualitative Perspective	8 August 2020	Preprint	[14]
(Tonnang et al., 2020)	COVID-19 Emergency public health and economic measures causal loops: A computable framework. In COVID-19	10 September 2020	Preprint	[15]
(Klement, 2020)	Systems Thinking About SARS-CoV-2	28 October 2020	Paper in a peer-reviewed journal	[13]
(Kontogiannis, 2021)	A qualitative model of patterns of resilience and vulnerability in responding to a pandemic outbreak with system dynamics	10 November 2020	Paper in a peer-reviewed journal	[4]
(Zięba, 2021)	How can systems thinking help us in the COVID-19 crisis?	8 June 2021	Paper in a peer-reviewed journal	[2]

To analyze the selected CLDs, we use both qualitative and quantitative methods. First, in Section 3.1 we discuss the research focus of the reviewed studies.

Second, in Section 3.2 we reveal commonly and rarely used concepts across the eight reviewed studies by identifying synonymic variables and computing simple statistics of the appearance of distinctively different notions across all CLDs.

Third, in Section 3.3 we analyze structural properties of the reviewed CLDs employing a number of approaches from the graph theory. Indeed, a CLD can be considered as a directed graph (a digraph) determined by its adjacency matrix $A = (a_{ij})$, $i, j = 1, \ldots, n$, where $a_{ij} = 1/a_{ij} = -1$ if component i makes a positive/negative impact on j and $a_{ij} = 0$ if i has no link into j; here, n is the total number of components in the considered system [27]. We compute and compare basic network statistics for the CLDs under review, including the number of nodes and links, as well as the average node degree, i.e., the average total number of the incoming and outgoing links associated with a node. We further analyze the dependence of links on the CLD size across the reviewed CLDs.

Fourth, in Section 3.4, for each CLD, we compute the statistics of the number of incoming and outgoing links associated with a node (in- and out-degree). Using the Frederic Vester's approach that was originally suggested in [28] and further developed by other authors in [29], we identify active components (drivers) and passive (most impacted) components of a CLD as nodes that have a high number of outgoing and incoming links, respectively. Components with high number of both link types are regarded as critical hubs in the corresponding CLDs.

Fifth, in Section 3.5, in each CLD we identify network motifs—basic microstructures, which can be considered as network building blocks. Following [30,31], we focus on (i) bidirectionality (a two-component feedback loop), (ii) multiple causes, (iii) multiple effects, (iv) an indirect effect, (v) a moderated effect, and (vi) three-component feedback loops (see Table 2). Motif (i) includes two nodes, while motifs (ii)–(vi) include three nodes. Furthermore, we run a conditional uniform random graph (CUG) test [30] to compare the prevalence of these motifs in the reviewed CLDs to their prevalence in the ensembles of random networks with the same number of nodes and edges (so-called "N, m" model family [32]) which is used as the null model.

Table 2. Network motifs used for analysis. Nodes highlighted with red depict impacting components, nodes highlighted with green depict impacted components. In the cases of bidirectionality and feedback loops, it is assumed that there is no dominant impact in any direction.

#	Motif Name	Motif Description (Following [33])	Motif Schematic View
(i)	Bidirectionality	A node impacts and is impacted by another adjacent node	
(ii)	Multiple causes	Two non-adjacent nodes impact another node, adjacent to both of them	
(iii)	Multiple effects	A node impacts two adjacent nodes which are non-adjacent between each other	
(iv)	Indirect effect	A node impacts a non-adjacent node through a third node	
(v)	Moderated effect	A node impacts an adjacent node both directly and through a third node	
(vi)	Feedback loop (3 components)	Three adjacent nodes impact each other in one direction, i.e., clockwise, or counterclockwise	

Sixth, in Section 3.6 we review how the eight CLDs were developed and presented in terms of the description of the design procedure, availability of lists of components, links and feedback loops, visualization, software implementation (source code), and methods employed for the CLD analysis. We selected these features as dimensions of good practice based on commonly used guidelines, e.g., [20] and our own practical experience.

3. Results

3.1. Research Focus

In terms of the research ambition, which the reviewed papers set for themselves, all eight papers share a similar approach that can be described as going "beyond health effects". This includes unraveling and visualizing the complexity and interconnectivity of different subsystems within the socioeconomic system, adding a transdisciplinary focus to COVID-19 policies, and identifying leverage points. Two papers specifically emphasize certain sectors, namely, refs. [2,26] concentrated on the role of media in the pandemic development coverage, and, in addition, ref. [2] considered the role of businesses behavior. We summarized the addressed research question in Table 3.

Table 3. Research questions addressed by the reviewed studies.

CLD ID	Research Question/Focus
(Wicher, 2020)	"I focused on the media and my role, as an individual, in the COVID-19."
(Bradley et al., 2020)	"< … > provide a framework to look beyond the chain of infection and better understand the multiple implications of decisions and (in)actions in face of such a complex situation involving many interconnected factors."
(Sahin et al., 2020)	"< … > visualise the complexity in managing the COVID-19 pandemic through a systems lens by identifying the interconnectivity between health, economic, social and environmental aspects."
(Bahri, 2020)	"< … > provide readers a qualitative analysis how the COVID-19 may affect our susceptible population, healthcare facilities and economy."
(Tonnang et al., 2020)	"< … > envision linkages between the elements of the contagion, healthcare, and the economy, and visualize key components that characterize the whole system."
(Klement, 2020)	"< … > try to identify and study system structures and causal loops of the problem at hand, integrating all relevant disciplines within an inter- and transdisciplinary approach."
(Kontogiannis, 2021)	"< … > unravel the nexus of social and institutional forces that affect the parameters of 'system dynamics' models < … >"; "< … > explore how CLDs, their modular blocks (i.e., system archetypes) and leverage points could be used to model < … > principles of resilience."
(Zięba, 2021)	"How do businesses respond to the prolonged exposure to the COVID-19 crisis? What kind of actions are they prone to undertake and what are the drivers of those actions?"

3.2. Common and Rare Components

In this section, we discuss similarities and differences between the components included in the CLDs by the authors of the reviewed studies. All eight CLDs accounted for the magnitude of the infection spread, and, in addition, studies [4,13,15,16] distinguished between the number of actually infected people and diagnosed cases ("Certified infections rates"[1] vs. "Infectious population" [4], "Number of positive tests" vs. "Infected population" [13], "Diagnosed" vs. "Infected" [15] and "Number of cases detected" vs. "Number of infectious people" [16]); studies [14,15] additionally distinguished between symptomatic and asymptomatic virus carriers. Six out of eight papers also separately accounted for the number of COVID-19 deaths. Three papers [4,14,15] use variables that are commonly included in SIR-type models, i.e., susceptible, recovered, and hospitalized populations [34].

The next most commonly included aspects across the eight CLDs are panic and/or fears (accounted for by six papers, i.e., "Panic and fear" [4,12], "Anxiety, panic and fear" [13], "Public outrage" [16], or just "Fears" [15]), as well as public awareness ("Alertness " [4], "Awareness campaign" [12], "Advisories and media reports" [15], "Effectiveness of public health risk communication" and "Public awareness" [16], "Situational awareness" [26]), business closures (lockdowns), unemployment (or "People out of work" [4]), impact on the healthcare system ("Hospital strain" and "Medical staff attrition" [4] "Health care worker load" [12], "Occupied health facilities" and "Shortage of health facilities" [14,15], "Impact on healthcare system" [26]), and social distancing (or "Avoidance of public space" [26])—each accounted for by five papers.

On the other hand, only two papers included the influence of the pandemic on the environmental issues, i.e., air pollution [12,13]. The former paper also accounted for the "Waste generation" and "CO_2 concentration". Social challenges such as (a lack of) "Trust within communities", "Crime and violence", and "Racism" [12], as well as the "Conflicts of interest" [13] appeared in only one paper, correspondingly. The role of vaccines was also highlighted only in two papers ("Development of vaccines", "Production with promising but not yet certified vaccine", and "Availability of vaccines" [4] and "Vaccination" [15]), while [4] is the only study which accounts for the role of research institutions ("Research institutes mobilisation"). Some issues that are generally considered important factors for the spread of COVID-19 and its impact, for example, social and economic inequality [35,36], are absent in all reviewed CLDs.

An exhaustive list of concepts used in all CLDs is provided in the Supplementary Material (Table S1).

3.3. Basic Network Properties of COVID-19 CLDs

Basic CLD network properties provide a simple indication of the system complexity. Table 4 presents the summary statistics that includes the number of nodes, the number of links, and the average node degree across the eight reviewed CLDs. CLDs vary significantly in terms of the number of nodes that they include: The smallest one (17 nodes) is by [2] who studied the business response to the COVID-19 crisis, and the largest one (78 nodes) is by [4], who analyzed the resilience of healthcare, government, social, and economic subsystems to the COVID-19 shock. Half of the reviewed CLDs have between 21 and 25 nodes, which corresponds to the commonly accepted standard [37].

Table 4. Comparative statistics of graph representations of the reviewed CLDs. The CLD highlighted in italics is an outlier in terms of average degree.

CLD ID	Nodes (n)	Links (l)	Average Degree ($\frac{2l}{n}$)
(Wicher, 2020)	21	37	3.52
(Bradley et al., 2020)	21	34	3.24
(Sahin et al., 2020)	*38*	*88*	*4.63*
(Bahri, 2020)	24	42	3.50
(Tonnang et al., 2020)	50	91	3.64
(Klement, 2020)	25	42	3.36
(Kontogiannis, 2021)	78	125	3.21
(Zięba, 2021)	17	32	3.77
Mean	34	61	3.61

Interestingly, across the reviewed CLDs, the number of links scales approximately linearly with the number of nodes. This can be seen in Figure 1 depicting the average node degree, which is twice the ratio of the number of links to the number of nodes. Excluding [12] as an outlier [38][2], we obtain that across the remaining seven CLDs, the average node degree is 3.46 ± 0.21. Such a narrow window of the average degree suggests that in most cases, the CLD developers in these seven studies regarded three to four links per element as an appropriate representation of the system's complexity in the context of their study. Study [12] involved a broader expert community into the design of their CLD, and this seems to have resulted in a more complex CLD with a much higher number of links and hence a higher average node degree—conceivably due to a larger heterogeneity of the views involved in the CLD construction [31].

3.4. Major Drivers and Most Impacted Components

Following Vester, in order to understand how a complex system can be managed, it is useful to identify active and passive components, as well as critical hubs in the corresponding CLD [28]. Active components have a substantial influence on other components of the system; changes in such components often trigger significant changes in the entire system, hence such components are often referred to as drivers. Passive (impacted) components tend to be sensitive to changes in other parts of the system. They can serve as indicators of the reaction of the system to a change, while they usually have a weak influence on the other components of the system. Critical hubs both strongly influence and are strongly influenced by other components of the considered system and often play an essential role in the formation of feedback loops [29].

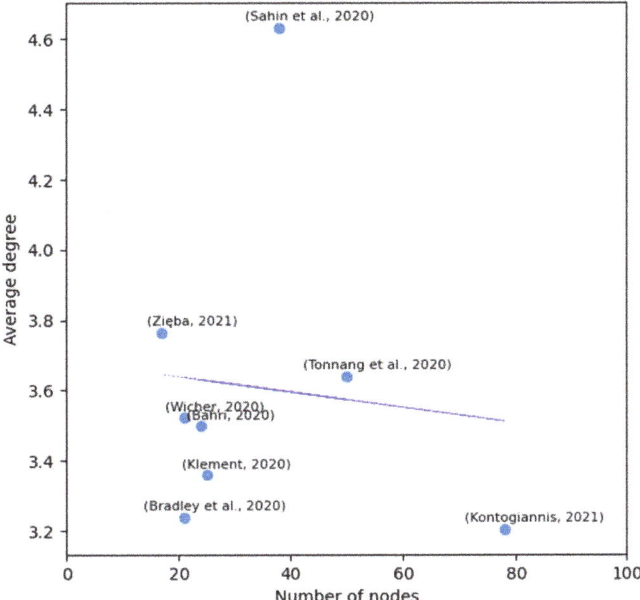

Figure 1. Average node degree across the reviewed CLDs as a function of their size, i.e., the number of nodes. The blue line represents the estimated linear trend excluding the outlier [12]. The slope is -0.004 with p-value 0.379, and hence the hypothesis that the average degree is independent on the network size cannot be rejected at the significance level at least 99.9%. The mean average degree value is 3.46 ± 0.21.

To identify the active and passive components in the reviewed CLDs, following the spirit of [28], for each reviewed CLD we obtain the in- and out-degree distributions, i.e., the observed frequencies of in- and out-degree values[3] and set a threshold which marks the highest distribution quantile. Here, we adopt the 10% right tail[4]. Those components, whose in-degree/out-degree is higher than the corresponding threshold value[5] are identified as candidate active/passive components. More details of the implementation of this procedure can be found in the Supplementary materials.

As components of a complex system typically both influence and are influenced by other components of the system, some may have both high out- and in-degrees. To deal with such cases, following [29], for each CLD component, we compute an active/passive quotient (APQ), i.e., the ratio of its out-degree to the in-degree. As "truly" active and passive components, for the further analysis, we select only those candidate active/passive components which have APQs greater/smaller than one. Furthermore, we determine critical hubs[6] as components that have a high product of out- and in-degrees. Table 5 summarizes the definitions used.

Table 5. Classification of system components following Vester.

	Active	Passive	Critical Hubs	
Out-degree	In the top decile	Any	Not in the top decile	In the top decile
In-degree	Any	In the top decile	Not in the top decile	In the top decile
Product of in-degree and out-degree	Any	Any	In the top decile	In the top decile
Active/passive quotient	>1	<1	Any	1

Across eight CLDs, two to seven components are classified as active[7]. A higher out-degree acts as an indicator of a higher importance of the component. In all reviewed papers, except [15], the magnitude of the infection spread expressed in terms of "Number of infected people" [26], "Infectious population" [4], "Number of positive tests" [13], "Confirmed cases" [12], "Perceived number of infectious people" [16], or, more generally, "Seriousness of the COVID crisis" [2] is a very important active component with the highest or second-highest out-degree. In both studies, where the infections are the second-ranked, the active component with the strongest influence is "(Policy) interventions" [12,13]. Government measures, "Lockdowns" and "Government imposed restrictions", are also important active components in [2,15], and both have the second highest out-degree in their corresponding CLDs.

"Health system" and "Centre for Disease Control" have the highest and the second-highest out-degree in [4,15] correspondingly.

An essential role of communication and media is reflected by the presence of "Public attention towards COVID-19" in [26], "Effectiveness of public health risk communication" in [16] and "Digital channels" in [15], and "Popularity of social media" in [2] among the active components in their corresponding CLDs. "Economic activities" [12] and "Economic pressure" [4] as well as "Unemployment" [15] also appear as important drivers in these papers. Finally, "Research Institutes Mobilization" imposes a strong influence on the entire system in [4].

Across the reviewed CLDs, two to nine components were classified as passive. The public perception of the pandemic is the most critical passive component in the CLDs of [2], expressed as "Perceived seriousness of COVID crisis"; of [12,13], expressed as "(Anxiety), panic and fear"; and of [16], expressed as "Public outrage". At the same time, "Chance of getting infected" has the highest in-degree in the CLD by [26]. The CLDs of [14] and [4,16] follow them, with "Infected droplets or surfaces", "Infection rate", and "Transmission events" having the second highest in-degree, correspondingly.

The most crucial passive component of the CLD by [4] is "Capacity to respond". At the same time, "Budget for fight the COVID-19" has the second-highest degree in the CLD by [26]. The most impacted component of [14]'s CLD is "Recovered population", while for [15], it is the "Isolated population".

Economic effects of the pandemic impact are reflected by the presence of "GDP loss fraction" [4], "Total demand" [15], and "Business closures" [12] among the passive components in the corresponding CLDs. Finally, influence of the pandemic on mental well-being is highlighted by the respective passive nodes in the CLDs of [13,14]. Interestingly, "Immune system" is active in the CLD of [15]'s CLD and passive in the CLD of [13].

"GDP" [14,15], "Situational awareness" [26], "Symptomatic population" [14], "Dead population" [14,15], "Isolated population" [14,15], "Hospitalized population" [4], "Mobilization of policies" [4], and "Vaccination" [15] can be defined as "critical hubs" as they impact and are impacted by many other components. For example, the numbers of symptomatic, hospitalized, and isolated people depend on how fast the virus spreads, but they also influence further contamination.

At the same time, our analysis of in- and out-degree distributions (Figure S1 in Supplementary Material) shows that six out of eight reviewed CLDs demonstrate a prevalence of transmitter variables, i.e., those with zero in-degree, over receiver variables, i.e., with zero out-degree, thus highlighting a shock character of the COVID-19 pandemic, which is considered as an external perturbation to a wider socioeconomic system.

A synthesis overview of active and passive components and critical hubs across all eight CLDs is schematically presented in Figure 2. These components are essential as "they are likely to have a bearing on a large number of issues and research questions" [39].

Figure 2. Summary of active and passive components of the systems. Concepts in red circles denote active components aggregated across the reviewed studies, concepts in green circles denote aggregated passive components, and concepts in yellow circles denote aggregated critical hubs.

3.5. Structural Complexity: Motifs

To measure how the reviewed CLDs reflect system complexity beyond the basic network statistics and to obtain insights regarding the degree to which a CLD represents a specific type of causality [30], we measure the prevalence of certain network motifs (listed earlier in Table 2) in each diagram.

For each of the reviewed CLDs, we generated 1000 random graphs with the same numbers of nodes and edges as the corresponding reviewed CLD. Then, for each of the six motif types, we calculated the expected number of motifs across the simulated random graphs and compared it with the actually observed number of motifs in the reviewed CLD. As a measure of motif prevalence, we chose a difference between the observed motif count and the corresponding expected value [31]. To be able to compare among CLDs which have different number of components and links, we standardize both the observed and expected numbers of motifs using the mean and standard deviation of the corresponding ensemble of random graphs (i.e., we compute z-scores) [31]. Therefore, the motif differences are measured in the number of standard deviations (Figure 3).

Our findings illustrate that all reviewed studies accounted for more bidirectional structures (feedback loops with two components) than might be expected. In this sense, the most prominent study is [15][8]. Six out of eight studies also had more three-component feedback loops than corresponding random networks. Four studies underrepresented and four studies overrepresented the moderated effects. At the same time, almost all studies demonstrated a lower prevalence of multiple effects (with the exception of [12]) and indirect effects (with the exception of [13]). In all studies, multiple causes were observed less frequently than in random networks.

Additionally, we find that there is no clear pattern between a motif's prevalence in CLDs and its size (measured by the number of components), as well as between a motif's prevalence and the date of its publication, i.e., more recent CLDs do not necessarily contain more complex causal structures than the early maps.

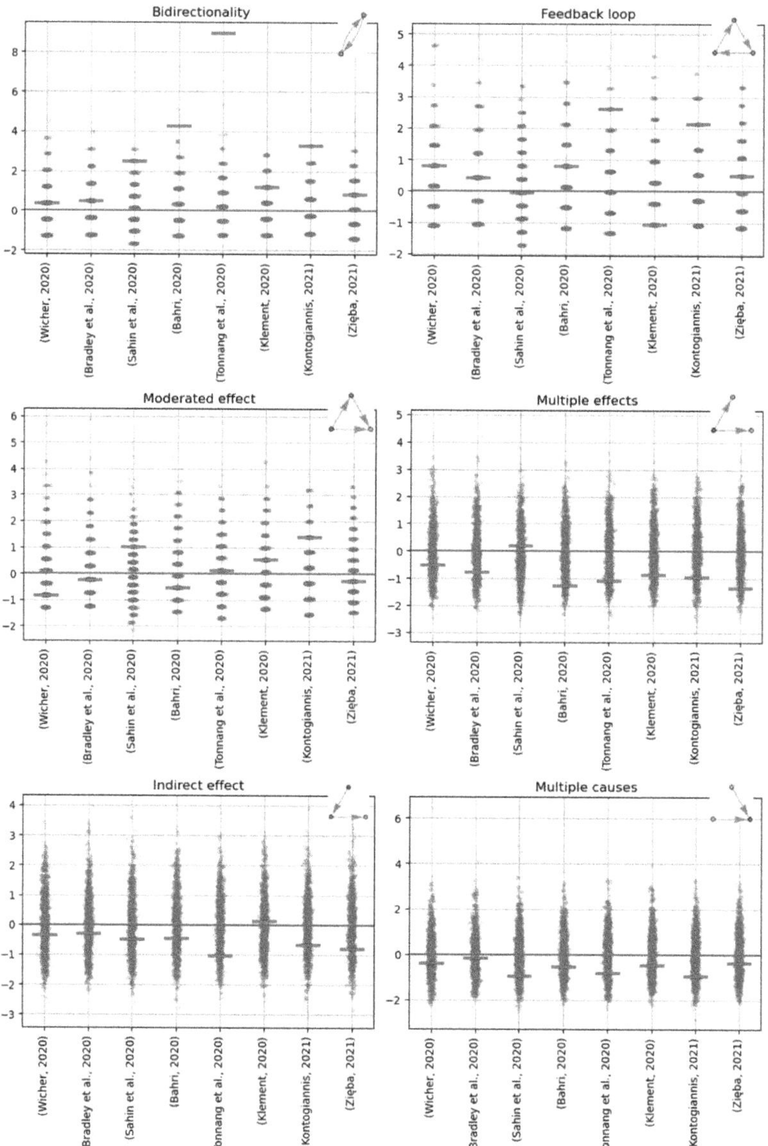

Figure 3. Grey dots represent the standard score[9] (z-score) of the number of motifs across 1000 realizations of the randomly generated graphs. The red mark depicts the actually observed indicator standardized in the same manner, so the red mark denotes the number of standard deviations by which the actually observed number of motifs differs from the mean of the distribution.

3.6. Good Practices of Creation and Visualization of CLDs

Here, we focus on the design procedure, availability of lists of components, links and feedback loops, visualization, software implementation (source code), and methods employed for the CLD analysis as important dimensions of good practice for developing and presenting CLDs. These features are selected based on commonly used guidelines, e.g., [20] and our own practical experience. Table 6 summarizes how these features are covered in the reviewed CLDs.

Table 6. Design and analysis features of the reviewed CLDs[10].

CLD ID	Design Procedure	List of Components, Links, and Feedback Loops	Visualization Features	Software Implementation	Analysis Methods
(Wicher, 2020)	Based on an analytical article	N/A	Feedback loops marked	N/A	Feedback loops
(Bradley et al., 2020)	N/A	N/A	The essential feedback loop is highlighted by color	Vensim	Feedback loops
(Sahin et al., 2020)	Based on expert workshops	Components	Subsystems highlighted by colored areas; feedback loops marked	Vensim	Feedback loops
(Bahri, 2020)	Based on data analysis and literature review	N/A	Separate CLDs of subsystems and archetypes; feedback loops marked	Vensim	Feedback loops, system archetypes
(Tonnang et al., 2020)	Formal description of the development process	Feedback loops	Subsystems highlighted by colored links; feedback loops marked	Vensim	Feedback loops
(Klement, 2020)	Built upon existing CLD	N/A	Subsystems highlighted by colored areas; feedback loops marked	N/A	Feedback loops
(Kontogiannis, 2021)	Built upon an existing SIR model and expert interviews	Feedback loops	Separate CLDs of archetypes; archetypes highlighted by color on the main CLD; feedback loops marked	Vensim	Feedback loops, system archetypes
(Zięba, 2021)	Based on "mental database, observation, and intuitive approach"	N/A	Feedback loops marked	Vensim	Feedback loops

An important prerequisite for the credibility, transparency, and replicability of a CLD is the description of its design procedure [40]. For example, the CLD presented in [12] was based on several expert workshops, which are briefly described in the paper's appendix (Appendix A) [12]. The CLD of [13] is based on this CLD. Expert workshops represent a useful source of unique knowledge and insights to address wicked problems [41]. Information on the workshop participants (can be anonymized) and other workshop details is useful for the readership to fully appreciate what the CLD represents. Using a coding procedure that formally translates participant statements into elements of a CLD can be recommended [42]. Another way to develop a CLD can be desk research. According to the descriptions provided, in six out of eight reviewed papers, the authors used their own mental models complemented by literature reviews to produce their CLDs.

We argue that a comprehensive description of the system's components (and ideally interconnections between them) and data sources that were used to inform them is important for the CLD validation. One of the reviewed studies provided such a description ([12], Appendix A). Some papers contain literature-based evidence for justifying some key interconnections between the components of their CLDs [2,4,13,14]. We believe that while general knowledge can often be sufficient to draw causal links, in some cases, especially when it involves a novel phenomenon, such as in the case of COVID-19, justifying links with the available evidence can greatly increase the CLD's credibility. For example, the CLD in [13] includes a "# of positive tests"→"COVID-19 deaths" link and no "Infected population"→"COVID-19 deaths" link. We find that this is not completely straightforward, and as readers, we would appreciate a justification of this choice.

Using colors and other ways to evince the CLD structure often improves its comprehensibility [43]. Among the eight reviewed papers, three papers mark subsystems within the considered systems: [12,13] highlight different subsystems using areas of different colors and [15] highlights links in different subsystems using different colors[11]. Both approaches seem to be helpful for better reading of the CLDs to which they were applied.

Furthermore, [4] uses different colors to highlight system archetypes, which are commonly encountered combinations of reinforcing and balancing feedback loops, often leading to an undesired behavior of the considered system [44,45][12]. This study, along with [14], also presents separate maps of system archetypes, which constitute the building blocks of the full CLD. This is considered a useful practice by [20].

At least six out of eight reviewed CLDs were implemented using Vensim software[13], which is a commonly used tool for designing CLDs (and systems dynamics models). None of the papers provided a source file of their model. For researchers who would wish to use a CLD developed by other authors, having such a file would save efforts on reproducing it, especially if the CLD is rather large.

The main methodological approach to the CLD analysis in all eight reviewed papers is based on the selection and discussion of several major feedback loops, which is a standard practice in the field. Feedback loops are essential to understand the behavior of a system's model and identify potential leverage points [46,47]. CLDs can contain thousands of feedback loops [48], however, it is often enough to discuss the most essential ones which are relevant to the problem at hand. Moreover, it has been proposed that the CLD dynamics are largely driven by a relatively small subset of feedback loops, namely, a Shortest Independent Loop Set (SILS) introduced by [27] is defined as a "set of shortest loops which are necessary to fully describe the feedback loop complexity of the model" [49].

For discussing feedback loops and their role, it is essential that each one that is referred to is depicted separately and/or described textually in a way that allows readers to clearly see all the constituting links [50]. For example, [4,15], two studies with the largest numbers of feedback loops discussed, provide tables listing loops and their interpretation, which helps readers to follow the authors' argumentation. Moreover, [4,26] give distinctive names to their loops, as suggested by [20], to increase the understanding of the function of each loop. In terms of feedback loops visualization, all studies except for [16] label the discussed feedback loops in their CLDs, which helps in following the corresponding discussion in the paper. The commonly accepted labelling style is to use either "RX", "BX", or just "X", where "X" is the identificatory (number) of the analyzed loop, "R" refers to a reinforcing loop, and "B" refers to a balancing loop. Two papers, [4,14], go deeper and, following [51], identify and analyze archetypal structures in CLDs which are indicative of system modularity [52].

Feedback loops and systemic archetypes enable a better understanding of some of the challenges which make the COVID-19 pandemic a wicked problem. This type of analysis shows the capability of systems thinking to be of particular use to make a step towards problem structuring [53]. Furthermore, it can help to identify leverage points, which can steer the systems towards a desired goal or away from an undesired behavior [47]. This is explicitly emphasized in two of the reviewed papers [4,12].

4. Discussion and Conclusions

In this paper, we analyzed eight studies aimed to illustrate the complexity and multidimensionality of the COVID-19 crisis using a practical tool of systems thinking—causal loop diagrams (CLDs). Here, we highlight some of the observations. First, we observed that the key components of the reviewed CLDs are consistent across all eight studies, however, different studies put different emphases on the main drivers and main affected components of the analyzed systems. This diversity of both drivers and affected variables supports the need for a transdisciplinary response to the pandemic [13].

The insights on common and rare components (Section 3.2), as well as on drivers and the most affected elements (Section 3.4) can be useful for future CLD developers and quantitative modelers to guide their research. For example, CLD analysts may decide to focus on gaps revealed in the existing CLDs, e.g., inequality, or they may choose to focus on the most important components to dig deeper into their dynamics and impacts. However, the scope of some CLDs could be quite narrow, and therefore, reusing concepts from them for a more general study should be done carefully.

Quantitative modeling and in particular systems dynamics (SD) modeling [21] can benefit from this review, as modelers can use the discussed CLDs a basis for their models. The author of [4] supports this point of view: "[CLDs] have the potential to be converted into Stock and Flow diagrams that allow quantification of results". For example, CLDs can be used to extend the traditional SIR-type system dynamic models to make them more realistic and useful for decision making [22]. The most essential system components identified in this review can guide the choice of variables in models.

Our insights in this part can also be useful for policy makers. The analysis of drivers (Section 3.4) can indicate candidate leverage points for the mitigation of the adverse consequences of COVID-19 and improve the resilience of the socioeconomic system to "provide a basis for effective response to the control of the pandemic" [4] and "bounce forward" from the shock caused by the pandemic [7]. The analysis of the most impacted components carried out in the same subsection can draw the attention of policy makers to areas where unintended and unwanted effects may be anticipated.

Second, we observed that the average number of links per node across the reviewed CLDs does not depend on the diagram size (Section 3.3). We proposed that this might be because the CLD developers regarded three to four links per node as an appropriate representation of complexity in their studies. This and other observations discussed in Section 3.3 can be useful for future developers of CLDs in the context of COVID-19 for benchmarking their models and planning their efforts and scope.

Third, we revealed a higher-than-expected prevalence of two- and three-component feedback loops in the reviewed CLDs (Section 3.5). This is different from the results obtained by [30], which found a low prevalence of these feedback structures in cognitive maps developed in the context of sustainable agriculture. This difference can be explained by the fact that the CLDs that we reviewed were developed by researchers familiar with systems thinking, which, according to [30], leads to a higher complexity of the developed cognitive models. Furthermore, in the same subsection and consistently with [30], almost all of the CLDs that we reviewed underrepresent "multiple effects" and "indirect effects" motifs, and they also underrepresent "multiple causes" motifs, which are, on the contrary, prevalent in [30]. The latter fact can probably be attributed to the novelty of the COVID-19 pandemic. Interestingly, while all authors discuss the feedback loops identified in their CLDs, none of them explicitly analyze multiple causes or effects for any components of the considered system[14]. This could be attributed to the fact that humans tend to perceive effects as more abstract and distant phenomena than causes, as suggested by the construal-level theory [54]. These observations can be useful for CLD developers for benchmarking their analysis as well as for researchers generally focusing on complexity and systems thinking.

Fourth, our observations made in Section 3.6 on good practices of development, presentation and analysis of CLDs can be helpful for future CLD developers. In terms of CLD development, we suggested that a detailed description of the design procedure enhances trust in the developed CLD. In terms of CLD presentation, highlighting meaningful subsystems of a large system helps reading a complex CLD. Finally, in terms of analysis, feedback loops and other smaller structures which constitute CLD building blocks such as archetypes and motifs can shine the light on the system complexity and help understand its behavior.

We conclude that despite the numerous recent calls to use systems thinking for addressing the complexity of the COVID-19 crisis, its practical applications are currently scarce; for example, [2] notes in this regard that "systems thinking approach to analyze the consequences of the COVID-19 outbreak is relatively novel and not extensively used". More recent studies generally do not contain more complex causal structures than the earlier ones. Therefore, we assume that they do not build upon the past models. Only one of the reviewed CLDs is explicitly based on another existing CLD. A plausible explanation of this fact is that CLDs are often developed for a specific purpose with a further aim to inform a more sophisticated model or analysis. However, we are not aware if any of the reviewed CLDs have been used for such a purpose up to the date of our writing.

We suggest (Section 3.6) that CLDs could benefit from a rigorous description of the development procedure and information sources used. This would improve their credibility and enable other researchers to enhance them further or conduct other types of analysis. Moreover, sharing the model source file can also be beneficial, especially since most of the reviewed maps showed consistency in the most important components and interactions and the degree of their complexity. Therefore, the reusability of CLDs could be key to enhance the efficiency of research efforts and/or to promote more advanced studies.

Being a useful systems-thinking tool, CLDs also have a series of limitations. As with every model, a CLD constitutes a major simplification of the considered real system. CLDs do not distinguish stocks and flows, which, along with the feedback structures, are the essential concepts in modeling systems behavior [20,46]. CLDs are inherently static and therefore cannot account for the dynamics of the modeled system, i.e., behavior over time [46][15], without being translated into a computer simulation model. CLDs invite users for a mental simulation, which, however, can be challenging even for relatively simple CLDs [20].

Notwithstanding these limitations, we argue that the reviewed papers demonstrate the power of systems thinking to inform a holistic picture of the pandemic's impact on a broader socioeconomic system. Indeed, CLDs are helpful for an initial exposition of the complexities brought about by COVID-19 for policymakers and the general public. They promote critical thinking [53] and show how deeply the pandemic affects all areas of human activity and that there is no easy "silver bullet" to solve this wicked problem [55], thus calling for a transdisciplinary approach. We suggest that building more comprehensive CLDs and having formal tools for their analysis [27,43,49] can further unleash the potential of systems thinking to inform decision making in circumstances of a wicked problem, such as the COVID-19 crisis—either as a standalone tool or as an input to more sophisticated models and analyses. As no single modelling approach can serve as a panacea for addressing a complex policy issue, CLDs should ideally be used in combination with other methods and models to provide reliable policy advice.

Supplementary Materials: The following are available online at https://www.mdpi.com/article/10.3390/systems9030065/s1. Table S1. Concepts used in the reviewed CLDs; Figure S1. Distributions of in- and out-degrees for each reviewed CLD.

Author Contributions: Conceptualization, N.S. and E.R.; methodology, N.S. and E.R.; formal analysis, N.S.; writing—original draft preparation, N.S.; writing—review and editing, E.R.; visualization, N.S. All authors have read and agreed to the published version of the manuscript.

Funding: This research received no external funding.

Data Availability Statement: Not applicable.

Acknowledgments: The authors acknowledge three anonymous reviewers for their detailed comments which helped to significantly improve the manuscript.

Conflicts of Interest: The authors declare no conflict of interest.

Notes

[1] Here, and in what follows, the names of CLD components in quotation marks are those originally used by their authors in the reviewed publications.

[2] The average node degree of the CLD from [12] constitutes 2.2 standard deviations from the mean of the ensemble of the eight CLDs under review. A threshold of two standard deviations is often considered enough to determine outliers in small-size samples [38].

[3] These can be computed as sums of the absolute values of rows and columns of the respective adjacency matrices.

[4] The 10% threshold is our choice to delimit a group of the most impacting/impacted components from the others. We show this in the distribution plots of the in- and out-degrees for each reviewed CLD (Supplementary Materials Figure S1).

[5] If several components with the same degree were divided by the top decile, all of them were considered.

[6] Vester originally classified all components with a high product of in- and out-degrees as critical, thus often including active and passive components. In this review, we emphasize the role of components, which are both systems drivers and indicators, but formally could not be classified as either active or passive. Formally we included components which either (i) have different in- and out-degrees less than top deciles or (ii) have equal in- and out-degrees in the top deciles, and, at the same time, have the product of in-degree and out-degree in the top deciles of the corresponding distributions of in- and out-degrees for each CLD.

[7] Vester also considers buffer components which have a low product of in- and out-degrees. These are beyond of scope of our analysis.

[7] The CLD by [14] does not have any active components fulfilling our criteria.

[8] Their CLD contained eight standard deviations more of the bidirectional structures than the random networks' mean.

[9] A linear transformation of raw data that provides that the mean and the variance of the distribution are 0 and 1, correspondingly. The standard score thus gives the number of standard deviations by which the actual data point is above or below the mean value.

[10] Table entries marked with "N/A" indicate that the corresponding aspect has been neither explicitly articulated by the authors or the reviewed studies nor it could be identified straightforward by the review authors.

[11] We assume that the authors of the reviewed CLDs have defined such subsystems a priori classifying components substantially, e.g., economic, social, healthcare, etc. However, it is also possible to recognize subsystems after a CLD has been developed, for example, using graph clustering methods.

[12] Usually four generic problem archetypes are specified [45]: (i) the underachievement, (ii) relative achievement, (iii) relative control, and (iv) out-of-control. While also being "building blocks" of CLDs containing few components, these are different to motifs discussed in Section 3.5.

[13] Three studies mention this explicitly, while the CLDs of three more studies have a typical visual appearance, which allowed us to attribute them to this software.

[14] Analysis of multiple causes and multiple effects (along with detection of feedback loops) for each component of a CLD can be performed using Vensim software (which was used to develop the majority of the reviewed CLDs and is commonly used for this purpose).

[15] Although six out of eight reviewed CLDs account for time delays for some of the links helping to qualitatively understand the speed of impact propagation, this still does not enable a formal analysis of the modeled systems' dynamics.

References

1. Dong, E.; Du, H.; Gardner, L. An interactive web-based dashboard to track COVID-19 in real time. *Lancet Infect. Dis.* **2020**, *20*, 533–534. [CrossRef]
2. Zięba, K. How can systems thinking help us in the COVID-19 crisis? *Knowl. Process Manag.* **2021**, 1–10. [CrossRef]
3. The OECD Forum Network towards a People-Centred, Inclusive, and Sustainable COVID-19 Recovery: OECD Launches the Centre on Well-Being, Inclusion, Sustainability and Equal Opportunity (WISE). Available online: https://www.oecd-forum.org/posts/towards-a-people-centred-inclusive-and-sustainable-covid-19-recovery-oecd-launches-the-centre-on-well-being-inclusion-sustainability-and-equal-opportunity-wise (accessed on 18 December 2020).
4. Kontogiannis, T. A qualitative model of patterns of resilience and vulnerability in responding to a pandemic outbreak with system dynamics. *Saf. Sci.* **2021**, *134*, 105077. [CrossRef]
5. Ilmola-Sheppard, L.; Strelkovskii, N.; Rovenskaya, E.; Abramzon, S.; Bar, R. A Systems Description of the National Well-Being System. Available online: http://pure.iiasa.ac.at/id/eprint/16318/ (accessed on 2 September 2021).
6. Ioannidis, J.P.A. Coronavirus disease 2019: The harms of exaggerated information and non-evidence-based measures. *Eur. J. Clin. Investig.* **2020**, *50*, e13222. [CrossRef] [PubMed]
7. Hynes, W.; Trump, B.; Love, P.; Linkov, I. Bouncing forward: A resilience approach to dealing with COVID-19 and future systemic shocks. *Environ. Syst. Decis.* **2020**, *40*, 174–184. [CrossRef]
8. Reynolds, S. COVID-19 Means Systems Thinking Is No Longer Optional. Available online: https://www.thinknpc.org/blog/covid-19-means-systems-thinking-is-no-longer-optional/ (accessed on 22 July 2021).
9. Haley, D.; Paucar-Caceres, A.; Schlindwein, S. A Critical Inquiry into the Value of Systems Thinking in the Time of COVID-19 Crisis. *Systems* **2021**, *9*, 13. [CrossRef]
10. Jackson, M.C. How We Understand "Complexity" Makes a Difference: Lessons from Critical Systems Thinking and the Covid-19 Pandemic in the UK. *Systems* **2020**, *8*, 52. [CrossRef]
11. Hassan, I.; Obaid, F.; Ahmed, R.; Abdelrahman, L.; Adam, S.; Adam, O.; Yousif, M.A.; Mohammed, K.; Kashif, T. A Systems Thinking approach for responding to the COVID-19 pandemic. *East. Mediterr. Health J.* **2020**, *26*, 872–876. [CrossRef]
12. Sahin, O.; Salim, H.; Suprun, E.; Richards, R.; MacAskill, S.; Heilgeist, S.; Rutherford, S.; Stewart, R.A.; Beal, C.D. Developing a Preliminary Causal Loop Diagram for Understanding the Wicked Complexity of the COVID-19 Pandemic. *Systems* **2020**, *8*, 20. [CrossRef]
13. Klement, R.J. Systems Thinking About SARS-CoV-2. *Front. Public Health* **2020**, *8*, 1–6. [CrossRef] [PubMed]
14. Bahri, M. The Nexus Impacts of the COVID-19: A Qualitative Perspective. *Preprints* **2020**, 2020050033. [CrossRef]
15. Tonnang, H.; Greenfield, J.; Mazzaferro, G.; Austin, C.C. COVID-19 Emergency Public Health and Economic Measures Causal Loops: A Computable Framework. *SSRN Electron. J.* **2020**. [CrossRef]
16. Bradley, D.T.; Mansouri, M.A.; Kee, F.; Garcia, L.M.T. A systems approach to preventing and responding to COVID-19. *EClinicalMedicine* **2020**, *21*, 100325. [CrossRef]
17. Taylor, I.; Masys, A.J. A System Dynamics Model of COVID-19 in Canada: A Case Study in Sensemaking. In *Sensemaking for Security*; Masys, A.J., Ed.; Springer: Cham, Switzerland, 2021; pp. 179–199.

18. Sy, C.; Ching, P.M.; San Juan, J.L.; Bernardo, E.; Miguel, A.; Mayol, A.P.; Culaba, A.; Ubando, A.; Mutuc, J.E. Systems Dynamics Modeling of Pandemic Influenza for Strategic Policy Development: A Simulation-Based Analysis of the COVID-19 Case. *Process Integr. Optim. Sustain.* **2021**, 1–14. [CrossRef]

19. Niwa, M.; Hara, Y.; Sengoku, S.; Kodama, K. Effectiveness of Social Measures against COVID-19 Outbreaks in Selected Japanese Regions Analyzed by System Dynamic Modeling. *Int. J. Environ. Res. Public Health* **2020**, *17*, 6238. [CrossRef]

20. Sterman, J.D. *Business Dynamics: System Thinking and Modeling for a Complex World*; Irwin McGraw-Hill: Boston, MA, USA, 2001.

21. Homer, J.; Oliva, R. Maps and models in system dynamics: A response to Coyle. *Syst. Dyn. Rev.* **2001**, *17*, 347–355. [CrossRef]

22. Kumar, A.; Priya, B.; Srivastava, S.K. Response to the COVID-19: Understanding implications of government lockdown policies. *J. Policy Model.* **2021**, *43*, 76–94. [CrossRef] [PubMed]

23. Adamu, A.A.; Jalo, R.I.; Habonimana, D.; Wiysonge, C.S. COVID-19 and routine childhood immunization in Africa: Leveraging systems thinking and implementation science to improve immunization system performance. *Int. J. Infect. Dis.* **2020**, *98*, 161–165. [CrossRef]

24. Shahabi, V.; Azar, A.; Faezy Razi, F.; Fallah Shams, M.F. Simulation of the effect of COVID-19 outbreak on the development of branchless banking in Iran: Case study of Resalat Qard–al-Hasan Bank. *Rev. Behav. Financ.* **2020**, *13*, 85–108. [CrossRef]

25. Zieba, K. How can Systems Thinking Help Us Handling the COVID-19 Crisis? In Proceedings of the 21st European Conference on Knowledge Management ECKM 2020, Coventry, UK, 3–4 September 2020.

26. Wicher, D. The COVID-19 Case as an Example of Systems Thinking Usage. Available online: https://agilejar.com/2020/03/a-great-example-of-systems-thinking-covid-19-case/ (accessed on 22 July 2021).

27. Oliva, R. Model structure analysis through graph theory: Partition heuristics and feedback structure decomposition. *Syst. Dyn. Rev.* **2004**, *20*, 313–336. [CrossRef]

28. Vester, F. *The Art of Interconnected Thinking*; MCB Publishing House: München, Germany, 2007.

29. Forgie, V.E.; van den Belt, M.; McDonald, G.W. Extending the Boundaries of Economics to Well-Being: An Interlinked Thinking Approach. In *Feedback Economics*; Cavana, R.Y., Dangerfield, B.C., Pavlov, O.V., Radzicki, M.J., Wheat, I.D., Eds.; Springer: Cham, Switzerland, 2021; pp. 521–544.

30. Levy, M.A.; Lubell, M.N.; McRoberts, N. The structure of mental models of sustainable agriculture. *Nat. Sustain.* **2018**, *1*, 413–420. [CrossRef]

31. Aminpour, P.; Gray, S.A.; Singer, A.; Scyphers, S.B.; Jetter, A.J.; Jordan, R.; Murphy, R.; Grabowski, J.H. The diversity bonus in pooling local knowledge about complex problems. *Proc. Natl. Acad. Sci. USA* **2021**, *118*, e2016887118. [CrossRef] [PubMed]

32. Butts, C.T. Social network analysis: A methodological introduction. *Asian J. Soc. Psychol.* **2008**, *11*, 13–41. [CrossRef]

33. Aminpour, P.; Schwermer, H.; Gray, S. The relationship between social identity and cognitive diversity in environmental stakeholders. *PsyArXiv* **2021**. [CrossRef]

34. Bärwolff, G. Mathematical Modeling and Simulation of the COVID-19 Pandemic. *Systems* **2020**, *8*, 24. [CrossRef]

35. Ahmed, F.; Ahmed, N.; Pissarides, C.; Stiglitz, J. Why inequality could spread COVID-19. *Lancet Public Health* **2020**, *5*, e240. [CrossRef]

36. Stok, F.M.; Bal, M.; Yerkes, M.A.; de Wit, J.B.F. Social Inequality and Solidarity in Times of COVID-19. *Int. J. Environ. Res. Public Health* **2021**, *18*, 6339. [CrossRef] [PubMed]

37. Betley, E.; Sterling, E.; Akabas, S.; Gray, S.; Sorensen, A.; Jordan, R.; Eustice, C. Modeling links between corn production and beef production in the United States: A systems thinking exercise using mental modeler. *Lessons Conserv.* **2021**, *11*, 26–32.

38. Wilcox, R.R. The Normal Curve and Outlier Detection. In *Fundamentals of Modern Statistical Methods*; Springer: New York, NY, USA, 2010; pp. 29–45.

39. Niemeijer, D.; de Groot, R.S. A conceptual framework for selecting environmental indicator sets. *Ecol. Indic.* **2008**, *8*, 14–25. [CrossRef]

40. Kim, H.; Andersen, D.F. Building confidence in causal maps generated from purposive text data: Mapping transcripts of the Federal Reserve. *Syst. Dyn. Rev.* **2012**, *28*, 311–328. [CrossRef]

41. Vennix, J.A.M. *Group Model Building: Facilitating Team Learning Using System Dynamics*; John Wiley & Sons Ltd.: Chichester, UK, 1996; ISBN 978-0-470-86668-9.

42. Eker, S.; Zimmermann, N. Using Textual Data in System Dynamics Model Conceptualization. *Systems* **2016**, *4*, 28. [CrossRef]

43. Stämpfli, A. A Domain-Specific Language to Process Causal Loop Diagrams with R. In *Operations Research Proceedings 2019. Selected Papers of the Annual International Conference of the German Operations Research Society (GOR), Dresden, Germany, 4–6 September 2019*; Neufeld, J.S., Buscher, U., Lasch, R., Möst, D., Schönberger, J., Eds.; Springer International Publishing: Berlin/Heidelberg, Germany, 2020; pp. 651–657.

44. Schoenenberger, L.; Schmid, A.; Tanase, R.; Beck, M.; Schwaninger, M. Structural Analysis of System Dynamics Models. *Simul. Model. Pract. Theory* **2021**, *110*, 102333. [CrossRef]

45. Wolstenholme, E.F. Towards the definition and use of a core set of archetypal structures in system dynamics. *Syst. Dyn. Rev.* **2003**, *19*, 7–26. [CrossRef]

46. Lane, D.C. The emergence and use of diagramming in system dynamics: A critical account. *Syst. Res. Behav. Sci.* **2008**, *25*, 3–23. [CrossRef]

47. Saleh, M.; Oliva, R.; Kampmann, C.E.; Davidsen, P.I. A comprehensive analytical approach for policy analysis of system dynamics models. *Eur. J. Oper. Res.* **2010**, *203*, 673–683. [CrossRef]

48. Kampmann, C.E. Feedback loop gains and system behavior (1996). *Syst. Dyn. Rev.* **2012**, *28*, 370–395. [CrossRef]
49. Schoenberg, W. LoopX: Visualizing and understanding the origins of dynamic model behavior. *arXiv* **2019**, arXiv:1909.01138.
50. Dhirasasna, N.; Sahin, O. A Multi-Methodology Approach to Creating a Causal Loop Diagram. *Systems* **2019**, *7*, 42. [CrossRef]
51. Senge, P.M. *The Fifth Discipline: The Art and Practice of the Learning Organization*; Doubleday: New York, NY, USA, 1990; ISBN 1368304091100.
52. Schoenenberger, L.; Schmid, A.; Schwaninger, M. Towards the algorithmic detection of archetypal structures in system dynamics. *Syst. Dyn. Rev.* **2015**, *31*, 66–85. [CrossRef]
53. Cavana, R.Y.; Mares, E.D. Integrating critical thinking and systems thinking: From premises to causal loops. *Syst. Dyn. Rev.* **2004**, *20*, 223–235. [CrossRef]
54. Trope, Y.; Liberman, N. Construal-level theory of psychological distance. *Psychol. Rev.* **2010**, *117*, 440–463. [CrossRef] [PubMed]
55. WHO. WHO Director-General's Opening Remarks at the Media Briefing on COVID-19. Available online: https://www.who.int/director-general/speeches/detail/who-director-general-s-opening-remarks-at-the-media-briefing-on-covid-19---3-august-2020 (accessed on 29 July 2021).

Article

Effectiveness of the Early Response to COVID-19: Data Analysis and Modelling

Edoardo Bertone [1,2,*], Martin Jason Luna Juncal [1], Rafaela Keiko Prado Umeno [3], Douglas Alves Peixoto [3], Khoi Nguyen [2,3] and Oz Sahin [2,3]

[1] School of Engineering and Built Environment, Griffith University, Southport, QLD 4222, Australia; martin.lunajuncal@griffithuni.edu.au
[2] Cities Research Institute, Griffith University, Southport, QLD 4222, Australia; k.nguyen@griffith.edu.au (K.N.); o.sahin@griffith.edu.au (O.S.)
[3] KODE Consulting, Parkwood, QLD 4214, Australia; rafaela.keiko@gmail.com (R.K.P.U.); kode.consulting2020@gmail.com (D.A.P.)
* Correspondence: e.bertone@griffith.edu.au; Tel.: +61-07555-28574

Received: 6 May 2020; Accepted: 16 June 2020; Published: 18 June 2020

Abstract: Governments around the world have introduced a number of stringent policies to try to contain COVID-19 outbreaks, but the relative importance of such measures, in comparison to the community response to these restrictions, the amount of testing conducted, and the interconnections between them, is not well understood yet. In this study, data were collected from numerous online sources, pre-processed and analysed, and a number of Bayesian Network models were developed, in an attempt to unpack such complexity. Results show that early, high-volume testing was the most crucial factor in successfully monitoring and controlling the outbreaks; when testing was low, early government and community responses were found to be both critical in predicting how rapidly cases and deaths grew in the first weeks of the outbreak. Results also highlight that in countries with low early test numbers, the undiagnosed cases could have been up to five times higher than the officially diagnosed cases. The conducted analysis and developed models can be refined in the future with more data and variables, to understand/model potential second waves of contagions.

Keywords: Bayesian Networks; COVID-19; pandemic; system thinking

---

1. Introduction

Based on official estimates, as of early May 2020, there are over 3,000,000 cases of COVID-19 worldwide with over a quarter of a million deaths. Such numbers are the result of a disease with a much higher (around 1%) fatality rate than a typical seasonal influenza [1]. Furthermore, it is caused by a virus (SARS-CoV-2) that is transmitted very efficiently, including by people who are only mildly ill or presymptomatic [2]. This high transmission ability by relatively healthy people makes it very difficult to contain the COVID-19 outbreak.

At the time of writing, most governments around the world have taken numerous actions in response to the COVID-19 pandemic to try to "flatten the curve", i.e., reduce the transmission rate in order to have a number of cases spread over a longer period of time. This is to avoid overcrowding hospitals over a short-term period, while also buying time to better prepare the country through more dedicated tools and facilities and better testing/tracing capabilities, with the end goal of "holding on" until a vaccine or an effective cure is developed. The magnitude and timing of government responses have varied remarkably. Countries such as Italy established a very heavy lockdown, with significant economic consequences, while other countries such as Sweden have adopted a lighter approach, with very limited restrictions and in turn, lower direct economic impacts. Of equal importance, is how society, and each individual, has reacted to the pandemic threat and adapted their lifestyle to the newly imposed rules or recommendations. Although it is proven that residents of heavily affected areas suffered from anxiety, stress, and other mental health issues [3], recent research also shows that the community response to COVID-19-related physical distancing measures is not necessarily high, and can vary considerably based, for instance, on a community's education and trust in science [4].

In synthesis, it is sensible to state that the effectiveness of a government response to the COVID-19 outbreak relies on its people, and that in turn, the community response is affected by the way their government handles the pandemic crisis, starting from how much and how consistently the importance of respecting restrictions is highlighted through different media outlets.

These complex interactions and the interconnectedness between government response, population response, COVID-19 cases, and deaths, and in turn, community mental health, country economy, climate, pollution, education system, population density, population age distribution, global travels, etc., makes understanding the causes and effects of the COVID-19 pandemic almost impossible with traditional approaches and with available data. Consequently, a systems thinking approach [5] is recommended to better quantify and understand such complex behaviours. This has been previously used by some authors to model complex multi-disciplinary problems [6,7]. A conceptual model, i.e., casual loop diagram, illustrating all the factors affecting the COVID-19 pandemic system, has been developed elsewhere [8]. Several of the aforementioned variables across the environmental-health-socio-economic subsystems are inherently difficult to numerically quantify; however, for some key variables, such as government and community responses, data currently exist through a number of online resources or other research studies. Therefore, by using a combination of traditional data-driven analyses and more complex systems approaches, such as Bayesian Networks [9], it was possible to model a small sub-system within the larger, overall COVID-19 pandemic network, to gain a better understanding and quantification of why certain countries have faster outbreaks and/or more deaths at this point in the pandemic crisis.

2. Results

2.1. Data Analysis Outputs

Firstly, Figure 1 illustrates a breakdown of countries hit the most by COVID-19 as of mid-April, based on how quickly the virus went out of control and caused several deaths. Specifically, it shows how many days passed before significant negative milestones, in terms of death counts, were reached. For every figure presented, the bullets represent the actual measurements whilst the lines are simply connecting the bullets for visual clarity.

Spain was the country that recorded the fastest spike in deaths, with only 31 days between recording the 100th case and 10,000 official deaths. Following Spain, Italy recorded the second quickest high death count, followed by the USA, France, and the UK, respectively. Following the 10,000 deaths milestone in Europe, both Italy and Spain were more successful than the UK and France in slowing down the death rate. Similarly, though the USA trajectory was the same, the exponential increase in deaths continued past the first 10,000 deaths, reaching the sad milestone of 20,000 deaths far quicker than any other country. In contrast, Germany recorded lower and later deaths at the beginning of the outbreak, as well as a slower increase in death count. Canada and Sweden had even an even slower and more delayed death count, while at the time of writing, Japan recorded only a few hundred deaths, which also started to accumulate well after the first few registered COVID-19 cases.

Figure 2 illustrates how prompt the overall response of different governments was in the early stages of their respective national COVID-19 outbreaks. A complete figure showing the overall time series based on normalized (Figure A1) and overall (Figures A2 and A3) number of cases, as per 10 April 2020 is provided in the Appendix A.

The lowest government action (GA—refer to Section 4.2) early scores were from Scandinavian governments, such as in Sweden and Norway. Spain, Italy, France, and Germany followed thereafter. The quickest countries to implement measures were Saudi Arabia, UAE, Japan, the USA, and Canada.

Australia had a moderate early response, though a constant stepwise introduction of new measures quickly made it the country with the highest GA score. Noticeably, these charts put the government action in perspective, based on the country population. Australia has a population which is about 13 times lower than the USA; hence, if the government action score was compared against the absolute number of cases, Australia would comparatively have a much prompter and earlier response, while the USA would plummet in this ranking (Figures A2 and A3). In Appendix A, the same charts for the Stringency Index [10,11] are presented for comparison purposes (Figures A4–A6). The trends are quite similar with the main differences being France, Italy, and Spain having comparatively a higher early SI than GA, while the USA, UAE, and Australia had lower SI scores in comparison to their respective GA results.

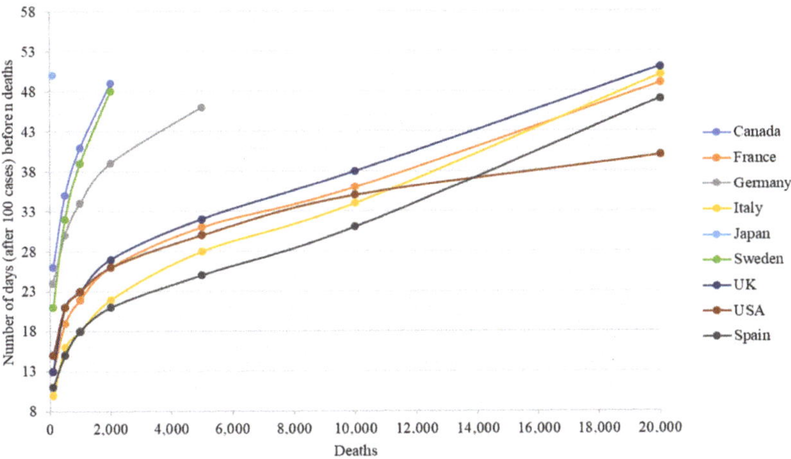

Figure 1. Number of days (starting from the day when 100 COVID-19 cases were recorded) before *n* COVID-19 deaths were recorded, where *n* is displayed along the *x*-axis (capped at 20,000 deaths).

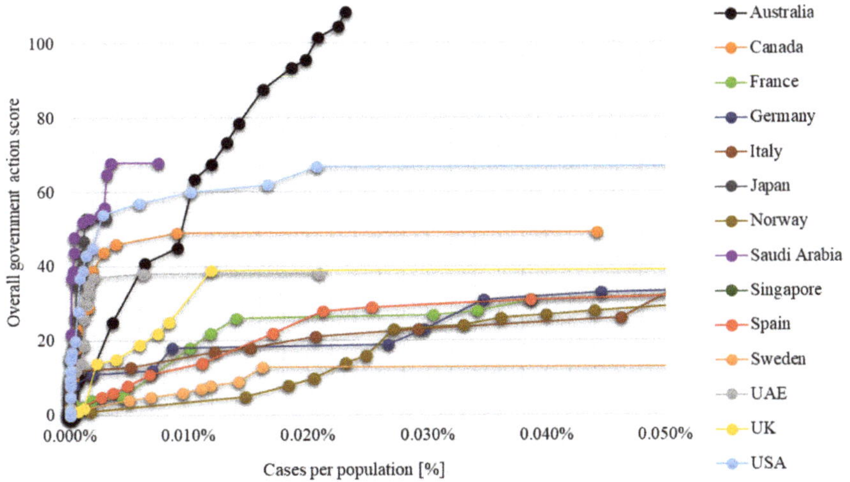

Figure 2. Overall government action score for different countries vs. recorded number of cases in proportion to country's population, limited to 0.05%.

Figure 3, in contrast, displays the calculated overall population action score (refer to Section 4.2), and its variation over time during the early stages of the outbreak.

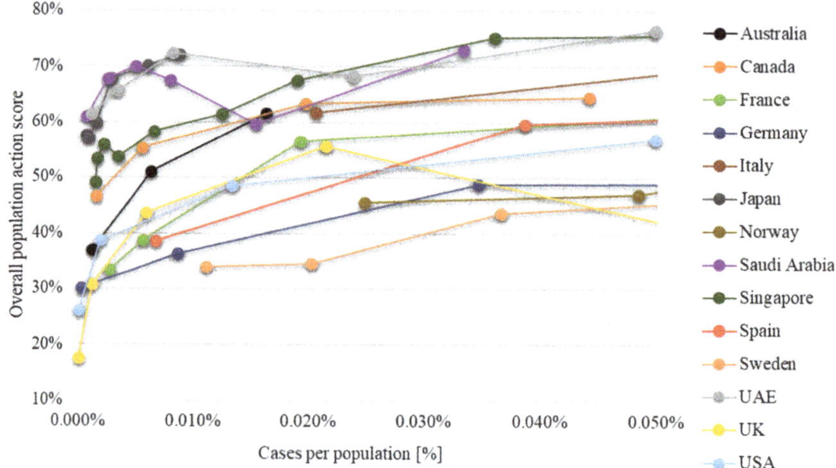

Figure 3. Overall population action score for different countries vs. recorded number of cases in proportion to country's population.

Both the UK and the USA started with very low scores, with values increasing over time to low-to-medium range values, with the UK score then decreasing again. Despite appearing to have an early and steep score increase, the large populations of the USA and UK compared to the other countries shown in Figure 3 highlights that their increase in population score was not particularly prompt when considering the absolute number of cases (Appendix A—Figure A7), but instead it occurred when several cases were already recorded. Germany and Sweden, although slightly better, recorded low scores and little improvement over time, while France started low but had a more significant improvement as cases increased. Canada, Italy, and Singapore had moderate initial scores, with improvements over time (Italy did not have early data as the outbreak in the country began before the survey study commenced). Japan, the UAE, and Saudi Arabia all had very high scores, although the latter showed a decrease over time.

Figure 4 displays the total number of reported tests performed over time in relation to the number of recorded cases.

A stark difference can be noticed between Australia, Germany, and Canada, and other countries such as the UK, USA, Sweden, Italy, and France. By the time 5000 cases were recorded in each country of the former group, approximately three times more tests were performed than by the countries in the latter group. Japan's testing numbers fall between the two aforementioned groups. Countries with no or limited data to more recent days (e.g., Spain or UAE) are not shown in Figure 4.

Relating to the above figure, Figure 5 displays the relationship between the amount of testing performed and the number of patients recovered in intensive care units (ICUs) at a specific point in time, when 5000 cases were officially recorded.

A non-linear negative relationship is evident, illustrating that countries with very low number of patients in ICUs, such as Australia, Germany, and the UAE, were, with the exception of the USA, those who performed the highest number of early tests. All countries recording high numbers of ICUs (e.g., Italy, Sweden, France) also performed the lowest number of early tests. As shown in later tables (Table 1) and Appendix charts (Figure A9), those countries with higher patients in ICUs and lower testing had a shorter time delay between the number of cases and number of deaths.

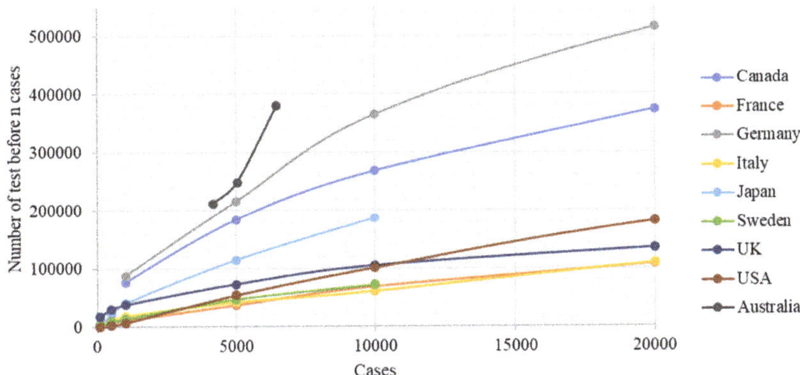

Figure 4. Number of COVID-19 tests performed by different nations before *n* cases were recorded, where *n* varies along the *x*-axis.

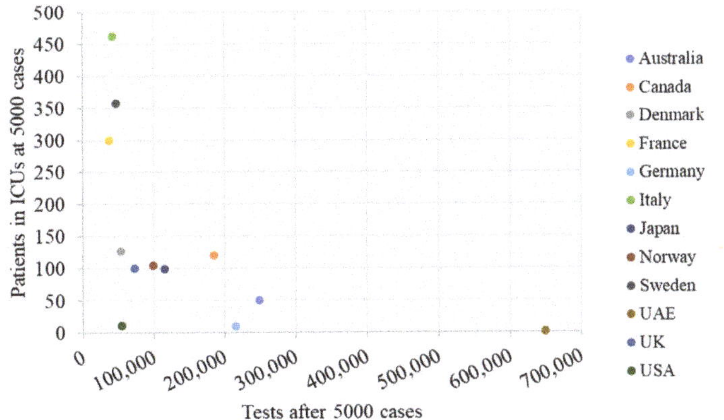

Figure 5. Relationship between the number of tests performed at the time 5000 COVID-19 cases were recorded, and the number of patients in ICUs at the time 5000 cases were recorded. Only countries with available data that recorded at least 5000 cases at the time of writing were included.

Table 1. Qualitative summary of the results and data for each analysed country.

Country	Days Before 10,000 Deaths	Early ICU	Early Gov Action	Early SI	Early Pop Action	Early Testing	Lag to Death (R^2)
Australia		L	H	M	L but +	H	7 (0.53)
Canada		M	H	H	M	H	14 (0.9)
France	36	H	L	M	L	L	6, 14 (~0.5)
Germany		L	L	L	L	H	12 (0.91)
Italy	34	H	L	H	H	L	6 (0.94)
Japan		M	L but +	M	H	M	10 (0.71)
Norway		M	M	L but +	L		12 (0.62)
Saudi Arabia			H	H	H but −		8 (0.69)
Spain	31		L	M	L		2 (0.94)
Sweden		H	L	L	L	L	7 (0.79)
UAE		L	M	H	H	H	8 (0.74)
UK	38	H	M	M	L	L	7 (0.92)
USA	35	L	H	M	L	L	7 (0.97)

"H" = high; "M" = medium; "L" = low; "+" = increasing with time; "−" = decreasing with time; blank = no data. "Lag to death" = number of days between number of cases and number of death providing the highest correlation. R^2 = coefficient of determination.

2.2. Bayesian Network Outputs

Figures 6 and 7 show the sensitivity analysis outputs of the Bayesian Network (BN) models, which were developed to predict the number of days before 5000 cases were reached (BN 1), and the number of days (starting from the day when 100 cases were recorded) before 1000 deaths were reached (BN 2). The numbers "0.02" and "0.05" relate to the % of cases (0.02% and 0.05%) against the total country population, as per Figures 2 and 3. In the figures, variables are ranked from those having the highest variance of beliefs (thus higher sensitivity) to those having the lowest one. Although the two BNs can be used to predict the two aforementioned variables, the focus in this section is on the sensitivity analysis since, rather than predicting, the main objective was to try to understand what factors cause a more (or less) rapid spread of the virus in the analysed countries. Sensitivity analysis made it possible to rank the different input variables in terms of their importance in affecting such spread, and thus they fulfil the purpose of identifying those population/government actions that most successfully helped reduce the diffusion rate of the virus.

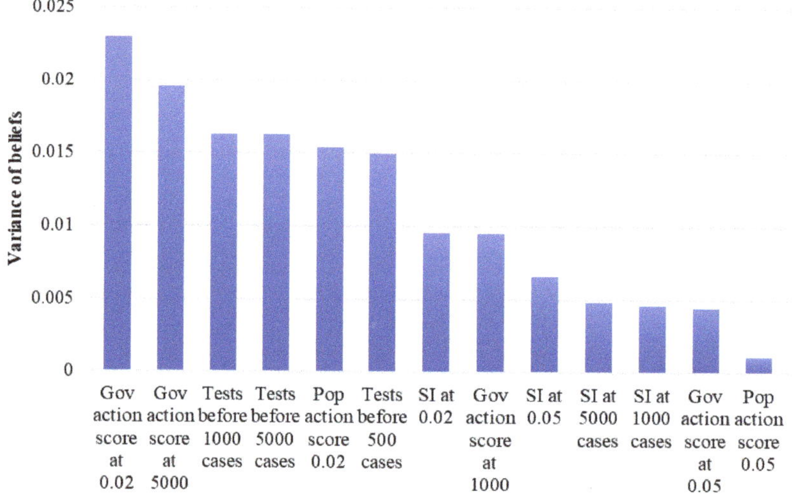

Figure 6. Sensitivity analysis outputs from BN 1, for the target node (days before 5000 cases). BN child nodes ordered from left to right based on variance of beliefs score.

It can be noticed that early (i.e., at 0.02% and at 5000 cases) government action is the most important factor in predicting the number of days before 5000 cases are recorded, since they are the two variables with the highest variance of beliefs. Conversely, the very early population action (0.02%) was much more important than population action at 0.05%, meaning that the way individuals behaved since the very beginning of the outbreak was crucial in establishing the transmission rate of the virus; however, the government response was even more crucial. Importantly, three out of the six most important variables were related to early number of tests. Finally, SI related variables were less important compared to the equivalent GA ones, providing an indication that the herein developed GA better captures the relevance of government actions in relation to the early transmission rate of the virus.

In relation to Figure 7, the three most important variables (i.e., with highest variance of beliefs) were all related to an early population response. Very early testing and stringency related variables followed, but with considerably lower importance (i.e., lower variance of beliefs). Overall, it appears that, while early testing amount emerged as important for predicting both early cases and early deaths, early government action was found to be significant in predicting/controlling early cases, while early population action was more important in predicting the early number of deaths.

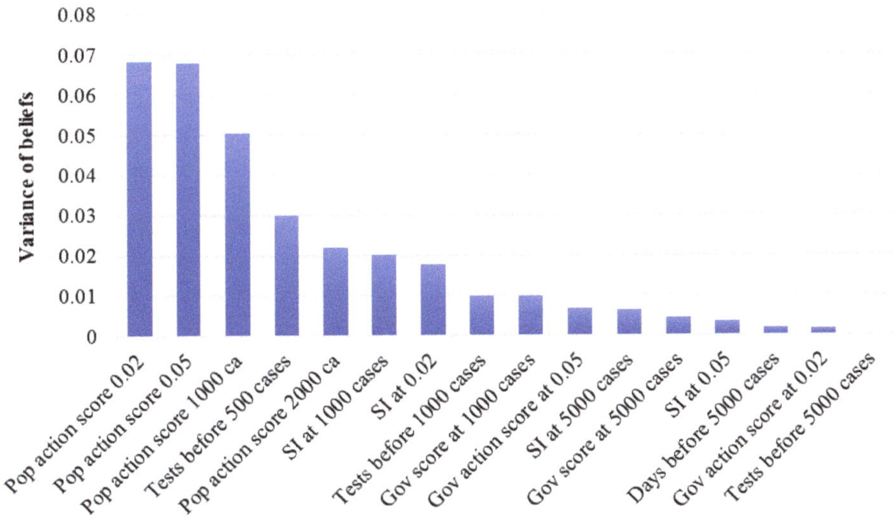

Figure 7. Sensitivity analysis outputs from BN 2, for the target node (days before 1000 deaths following 100 cases). BN child nodes ordered from left to right based on variance of beliefs score.

3. Discussion

The table below (Table 1) qualitatively summarises the data presented in the Results section for each country. At the time of writing, only France, Italy, Spain, the UK, and USA had reached 10,000 COVID-19 related deaths. Interestingly, all of them have a very low amount of early testing performed, as well as poor government or population responses (or both). Lack of early high testing numbers seems to emerge as a crucial, missing action that resulted in an uncontrolled, rapid spread of the virus. The more early, timely, and targeted tests, the more people with mild symptoms could be identified, thus isolating them before they could spread the virus further. Unlike previous outbreaks such as Ebola, the lethality of COVID-19 is significantly lower and usually results in mild to no symptoms for most infected people. As a result, it is much more difficult to identify and control. Therefore, it is logical that a lack of appropriate amounts of testing in the early days of the outbreak did not allow those countries to contain the virus. The under-detection of infected patients is clear from the significantly higher number of patients in ICUs, given the same overall number of cases diagnosed. Early studies [12] showed that approximately 4% of symptomatic patients in different Asian countries had to go through the ICU; in Italy, once the number of daily tests was finally boosted throughout April, the proportion of ICU patients compared to the total active cases followed a decreasing trend, from 4% towards 2%. With statistical studies and early serological surveys showing that the true number of infected, and particularly asymptomatic patients being significantly higher than reported through tests [13], it is safe to say that the number of patients in ICU would represent much less than 2% of the real, total amount of infected people. Regardless, even if 2% is taken as a reference, given 462 patients had already recovered in Italian ICUs at the time 5000 cases were detected, this translates to a more realistic figure of infected patients being over 23,000, which is almost five times higher than the official 5000 recorded cases, resulting from only 42,000 tests. With over 18,000 untested cases and the vast majority of them most likely having mild or no symptoms, while also being able to move around for several days before the first major lockdown rules were established on 9 March 2020, the Italian COVID-19 outbreak was already well underway and unnoticed before significant action could be taken. Our results from Figure 6 illustrate that early government action is crucial in controlling the speed of the outbreak, especially if early tests were limited: this is sensible since early government responsiveness could have helped in Italy and other countries to better control the untested, infected citizens who likely contributed significantly to the spread. Instead, the consequence was an early overcrowding of hospitals, leading to an extremely high number of deaths. The shorter lag between the time series of daily cases and daily deaths supports this hypothesis, since it seems that due to overcrowding and unpreparedness, hospitalised and ICU patients had less support and lower chances of survival, with only a week passing between the peaks in cases and peaks in deaths. This is similar for the other hard-hit European countries. Our findings from Figure 7 point at the early response of the population as critical in limiting the number of deaths within the first few weeks of the outbreak; with the death toll being a more robust measure of the diffusion of the virus, compared to the number of cases (biased and proportional to the number of tests performed), the citizen's risk perception of the virus, and the way they abide to the restrictions and rules established by their respective governments emerged as crucial indicators of the severity of the early spread of the COVID-19 outbreaks.

Interestingly, the population's response is itself affected by the government response; countries such as the UK and USA, whose initial public messages seemed to downplay the severity of the COVID-19 emergency, had a low initial population response (Figure 3), with citizens not feeling particularly worried and in turn, not practicing increased personal hygiene or wearing face masks. A systems thinking approach is crucial for understanding all these interconnections; the proposed BN models provide a first step in this direction. With a greater quantity of more reliable data becoming available, these models can be improved and refined over time.

Germany is the only large European country that successfully contained the outbreak from a death toll perspective; despite limited government action aside testing, the very high number of early tests allowed them to more effectively control the outbreaks and individual clusters, since a higher number of infected people with mild symptoms were detected and isolated. The delay between recorded number of cases and recorded number of deaths for this country is two weeks, resulting from an early testing response, an excellent healthcare system and a younger average population than Italy [14]. All other countries with a high amount of early tests, such as Australia and Canada, were able to control the outbreak and, in the case of Australia, completely "flatten the curve" at the time of writing, thus managing to contain the number of cases and deaths, as it can be seen from the data we collected and analysed. Hence, it seems that early government action becomes crucial only if early testing was limited (leading to several untested, infected people, free to spread the virus in their communities if no strict rules are imposed). This seems to be validated by the example of South Korea, which is not analysed in this study due to partial lack of necessary data, where government measures were limited, but the country managed to control the outbreak and flatten the curve by establishing an aggressive testing and contact tracing regime, while also enforcing quarantine policies [15].

An interesting case is provided by Sweden. Sweden is well-known for having adopted a "relaxed" approach to dealing with the COVID-19 pandemic [16]. In order to avoid catastrophic economic consequences, they did not impose a full lockdown, with very mild restrictions put in place instead. Although the government view suggested that they would rely on the citizens to do the right thing, the surveys highlight that the population response was instead quite poor. This unexpected response is then aggravated by a very low number of early tests performed. Although the number of cases and deaths seem to be relatively low, they are comparatively much higher than neighbouring Scandinavian countries such as Norway and Finland, and still rising at the time of writing. The high number of early patients in ICUs, coupled with low testing, seems to point at a higher number of actual infected cases (as high as 13,000 undetected) which, with more delay compared to other European countries, is now causing a gradual spread.

There are obviously several other factors that might play a role in the spread of COVID-19, which were not analysed here due to the lack of data or scientific evidence, such as population density and age distribution, or climate [17,18]. The developed BNs provide a way to quantify the importance of the analysed factors and provide a probabilistic prediction of the speed of the spread of COVID-19. Once more research consistently highlights the importance of other factors; these and related data can be easily incorporated in the BN structure and algorithms to reduce uncertainty.

4. Materials and Methods

4.1. Data Collection

The data used in this study were collected from numerous freely available online sources. Data for number of cases, deaths, cases in serious conditions, and tests were collected from Worldometers.info. As the website states, Worldometer is run by a team of international researchers, developers, and volunteers without any political, governmental, or corporate affiliation. With regards to COVID-19 data, the data are collected regularly from official Government sources or reliable media outlets. The data is then validated by a team of researchers before being published online. The data were collected for the February 1–April 16 period, i.e., from the onset of the outbreak to the time where the exponential trajectory of many European countries started to slow down, and in turn, where the effects of certain government measures became evident. The data were collected during a specific day, and when time series versions were not available, we accessed archived versions of the Worldometer COVID-19 main webpage through websites such as web.archive.org. Data about number of COVID-19 tests were also collected, or validated against, data from ourworldindata.org. Population behavioural response data were collected from a publicly available dataset, illustrating the results of a research work, conducted by YouGov and the Imperial College London—over population samples from 29 different countries. The data is in the form of weekly survey responses to 18 questions in relation to COVID-19 [19]. All the available data up to April 16 were collected. Regarding the quantification of the response of different governments, a full database of descriptive information consisting of a range of government actions around the world was available and downloaded from the ACAPS Government Measure Dataset [20] and other available online sources, as of 2 April 2020.

4.2. Data Pre-Processing and Analysis

The government action data were grouped into one of the following categories: visa restrictions, additional health documents required on arrival, border closure, domestic travel restrictions, emergency administrative structures, economic measures, restriction enforcement and surveillance, health protection, health screenings in airports and borders, lockdown, limit public gatherings, public services closures, psychological support, quarantine policies, schools closure, state of emergency declared, strengthening public health system, and testing policy. Once the category was chosen, each intervention was then assigned a degree of severity, on a scale from 1 to 4 (maximum). For instance, discouraging certain travel types was classified as a visa restriction Level 1, while a complete travel ban was denoted as Level 4. In addition, since certain measures were location-specific, this was incorporated within the severity degree. For instance, a strict lockdown on a specific region was given a score of 2, similar to a mild lockdown that was enforced over an entire nation. A strict, nationwide lockdown would be a Level 3 out of 3. Subsequently, since some of the categories could be cross-correlated, 5 wider groups were created by summing the scores of the relevant categories. These 5 groups were: (1) Political (e.g., special structures and enforcement groups); (2) coping/curing (e.g., testing measures, health facilities); (3) external control (e.g., border closures, visa restrictions); (4) internal control (e.g., lockdown, no public gatherings, school closures); and (5) socio-economic (e.g., government support to unemployed). Finally, an overall "government action score" GA was also calculated for each country by summing all the five individual scores. All such scores were calculated over the entire analysed time period, daily. These scores were then analysed over time, and in relation to the number of normalised cases (i.e., in relation to the nation's population).

Similar indexes, at the time of writing, have been developed elsewhere such as the Stringency Index (SI), which relies on a slightly different set of government response indicators and aggregated indices [10,11]. Such SI was also analysed in a similar fashion to the herein developed government action score for comparison purposes; this was done towards the end of our research work, hence SI data were collected from [10] and analysed as of 2 May 2020. SI-related variables were included in the developed Bayesian Networks, as explained in Section 4.3.

With regards to the population behavioural response, an overall "Population Action Score" was also calculated by averaging the survey results to a number of relevant questions, specifically: % of people (1) with fear of catching the virus, (2) avoiding crowds, (3) wearing a face mask, (4) practicing improved personal hygiene, and (5) not touching objects outside. This overall score was also analysed against the normalised number of cases. Analyses on results for individual survey questions, not shown here, was also conducted before calculating the overall Population Action Score.

Visual data inspection and time series analyses were performed to check the rapidity of the spread of the virus, by calculating the number of days before a country reached certain milestones with regards to cases and deaths. For these days, the number of tests conducted, as well as the number of patients in ICUs, was collected when available and used to understand their relationships with the rate of the virus spread, along with the other data. Furthermore, the time series for number of cases and the time series for number of deaths were analysed, and the time delay (lag) between them, which maximised the coefficient of determination (R^2), was also calculated for each country. Twenty-nine countries were initially selected, though not all were fully or partially analysed, due to either missing data or due to having, at the time of writing, limited cases and deaths. Figure A9 shows the results for the final set of the 17 countries analysed where data availability was sufficient at the time of writing.

4.3. Model Development and Application

Following the outcomes of the data analysis, a number of candidate input variables were selected and used to develop data-driven naïve and Tree-Augmented Naïve (TAN) Bayesian Network models, to try to predict critical variables linked to an early spread of the virus, specifically (1) Number of days before 5000 cases were reached (BN 1); and (2) number of days (after 100 cases) before 1000 deaths were reached (BN 2). Bayesian Networks rely on Bayesian theory, which in turn implies that the Bayes' theorem [21] can be used to infer or also update the degree of 'belief' given new information. They are made of variables called "nodes"; each variable is discretised in a number of "states". An "arc" connects a "parent" node to a "child" node. The relationship between a child node and its parent node(s) is quantified through a so-called conditional probability table (CPT). Populating CPTs can be performed based on either numerical, or qualitative (e.g., expert opinion), data. Bayesian Networks are an increasingly popular probabilistic modelling approach, which is well suited when only limited, uncertain, and incomplete data are available, such as for this case [9,22,23]. Figure 8 illustrates the structure of one of the developed BN.

Node discretisation and conditional probability table elicitation was performed and optimised from the data. In general, a naïve BN consists of only one parent node with multiple child nodes; more theoretical details can be found elsewhere, e.g., [9,24]; a TAN BN instead relaxes the strong independence assumption between all the child nodes given the parent [25], and thus arcs between child nodes are added. This can be noticed in Figure 8, where obvious links were added between those child nodes logically dependant on each other (from a temporal point of view). The final TAN structures were preferred to the naïve BN structures as typically they perform better [26] and they add logical connections between, in this case, temporally related nodes. The software used was Netica 5.22 32 bit (Norsys Software Corp, Vancouver, BC, Canada); the Netica API is available for download from their website [27]. Sensitivity analyses were completed using the in-built Netica algorithms; specifically, the sensitivity of different nodes was quantified by the "variance of node beliefs" (formerly named "quadratic score" in older Netica versions): this is defined as the expected change, squared, of the beliefs of the target node, taken over all of its states, due to a finding at the node in consideration [28]. It varies between 0 and 1, where 0 would represent that the target node is independent of the node in consideration, while the higher the value, the more sensitive the target node is to the node in consideration.

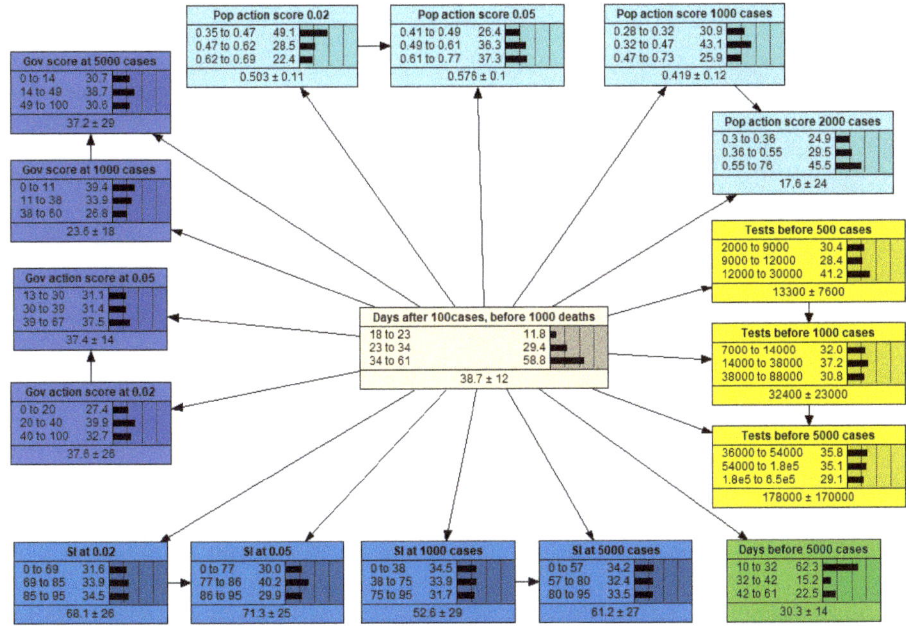

Figure 8. TAN Bayesian Network structure for "days after 100 cases before 100 deaths" (BN 2). Blue nodes: Government action score variables. Dark blue nodes: Stringency Index variables. Light green nodes: Cases variables. Yellow nodes: Testing variables. Light blue nodes: Population action score variables. "at 0.02" or "at 0.05": the day that 0.02% or 0.05% of the country's population tested positive.

5. Conclusions

A number of data analysis and modelling approaches were deployed to understand the importance and effectiveness of early government and population responses to COVID-19 outbreaks in several countries. Out of all the data and variables considered, high numbers of early tests emerged as the most crucial measure to control the transmission rate, as greater numbers of earlier tests lowered the number of undiagnosed and non-isolated cases. We estimated that countries with a low initial testing regime, such as Italy, might have had five times more actual cases than what was diagnosed. Following testing, early effective government responses were strongly related to slowing down the number of new recorded cases. Finally, the level of early population response, which in many ways is related to the type of government approach, was strongly related to the number of early deaths, which is a more reliable indicator of the spread of the virus. These conclusions point at the equally important contribution of a rapid government response and an early population-based behavioural change to abide with the new rules and health recommendations, which, in conjunction with aggressive early testing policies, assisted in controlling and managing early COVID-19 outbreaks. Due to the interconnectedness of the study's variables, a systems thinking approach is recommended for future studies to capture the inherent complexities of such a multidisciplinary problem. The developed Bayesian Network models have the ability to capture some of this complexity and related uncertainty, and can be refined and expanded to include more variables and data in the future, when they become available, to gain an even better understanding and improvement of the early management COVID-19 outbreaks. This will be of crucial importance as governments have started to lift some of the restrictions and are preparing for a potential "second wave" of infections.

Author Contributions: Conceptualization, E.B.; methodology, E.B., O.S.; software, K.N., D.A.P.; validation, E.B., M.J.L.J.; formal analysis, E.B., M.J.L.J., and R.K.P.U.; data curation, E.B., M.J.L.J., and R.K.P.U.; writing—original draft preparation, E.B.; writing—review and editing, M.J.L.J., R.K.P.U., O.S., K.N., and D.A.P. All authors have read and agreed to the published version of the manuscript.

Funding: This research received no external funding.

Acknowledgments: We are thankful to all the contributors to the online resources used for this research, and to KODE Consulting for providing a prototype version of the BN software SAMBA for validation purposes.

Conflicts of Interest: The authors declare no conflict of interest.

Appendix A

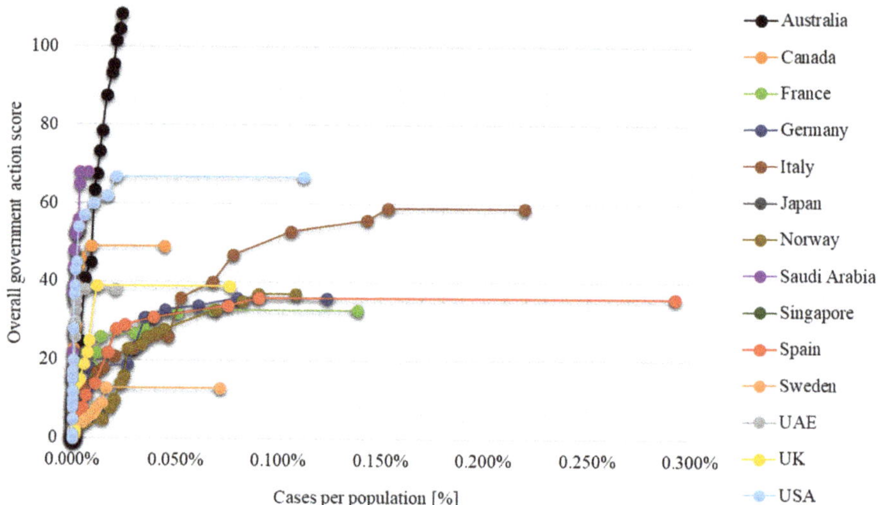

Figure A1. Overall government action score for different countries vs. recorded number of cases in proportion to country's population.

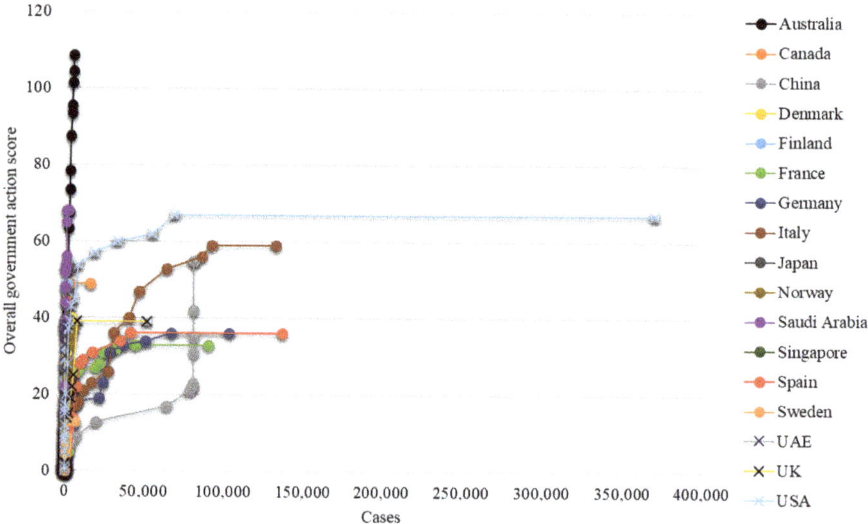

Figure A2. Overall government action score for different countries vs. recorded number of cases.

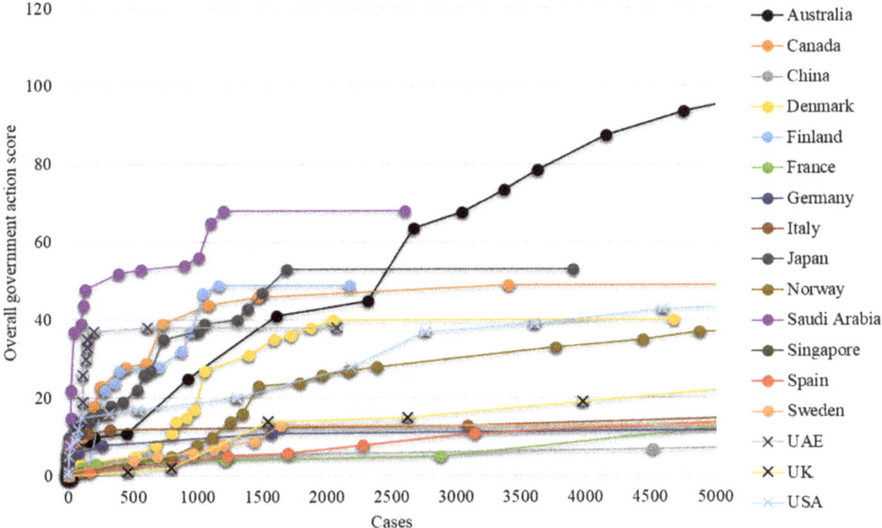

Figure A3. Overall government action score for different countries vs. recorded number of cases (limited to 5000).

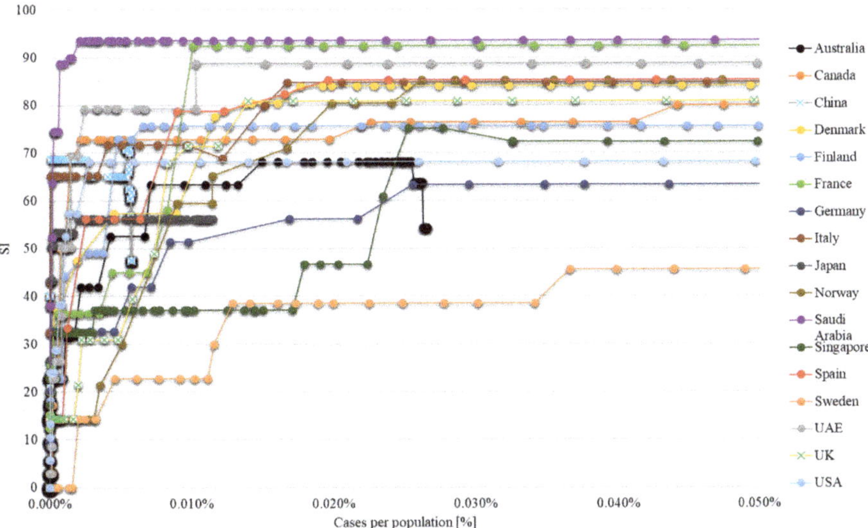

Figure A4. Stringency Index for different countries vs. recorded number of cases in proportion to country's population (limited to 0.05%). Updated 2 May 2020.

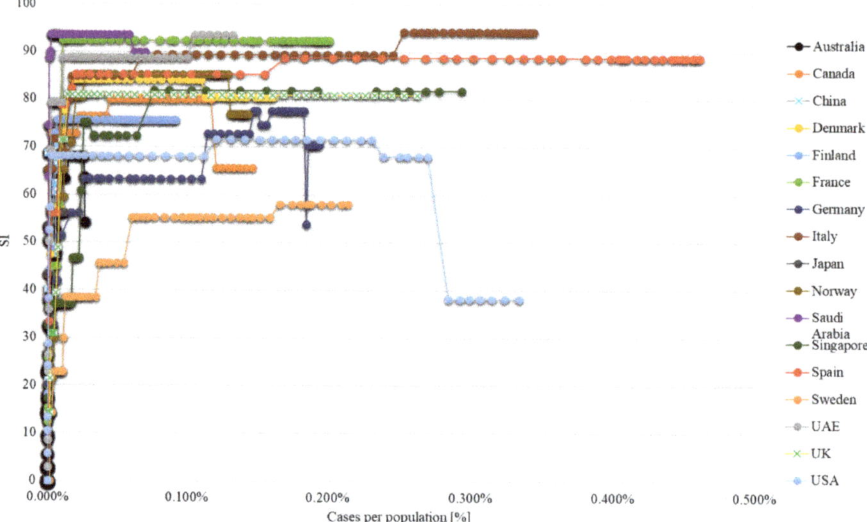

Figure A5. Stringency Index for different countries vs. recorded number of cases in proportion to country's population. Updated 2 May 2020.

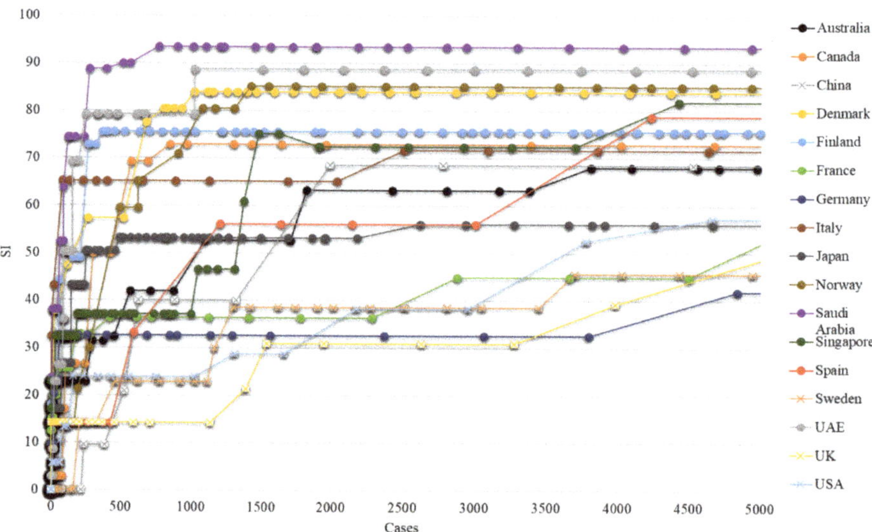

Figure A6. Stringency Index for different countries vs. recorded number of cases (limited to 5000). Updated 2 May 2020.

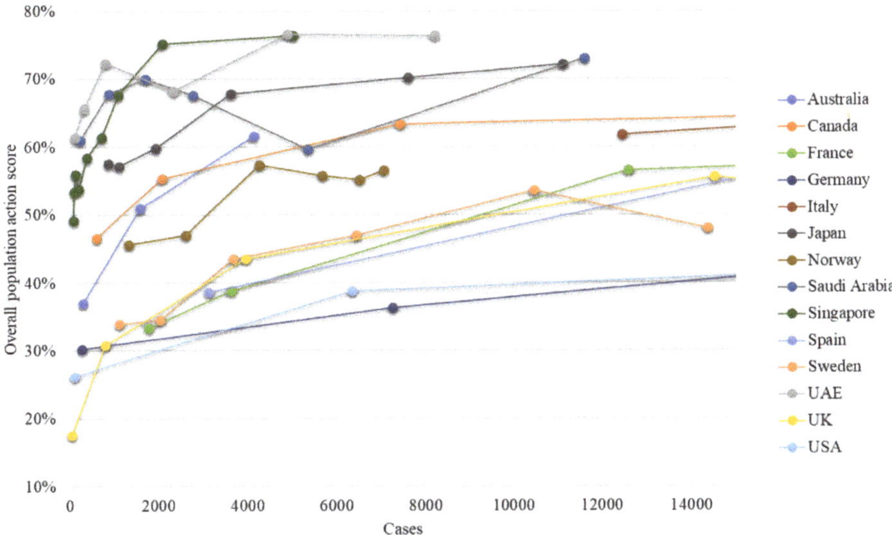

Figure A7. Overall population action score for different countries vs. recorded number of cases (limited at 15,000).

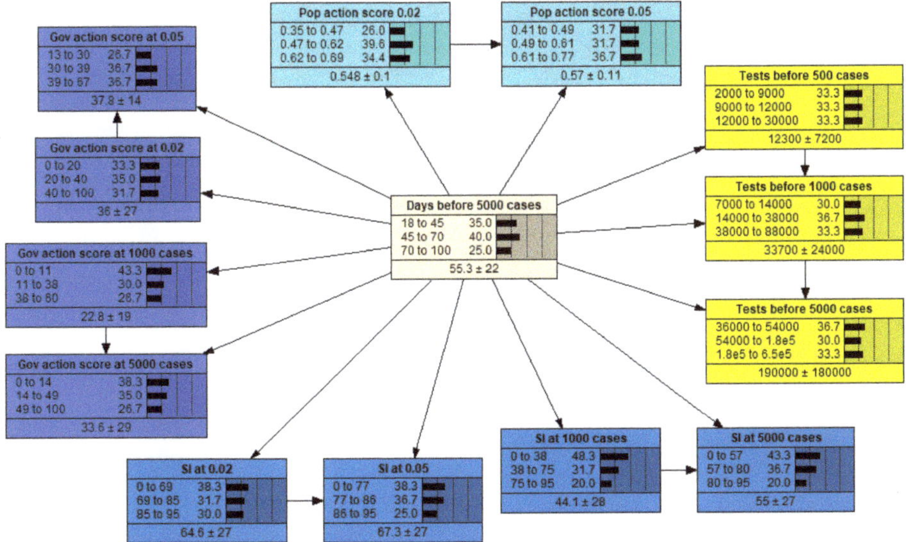

Figure A8. Bayesian Network structure for "days before 5000 cases". Blue nodes: Government action score variables. Dark blue nodes: Stringency Index variables. Yellow nodes: Testing variables. Light blue nodes: Population action score variables. "at 0.02" or "at 0.05": the day that 0.02% or 0.05% of the country's population tested positive.

	Country								
n	Australia	Canada	China	Denmark	Finland	France	Germany	Italy	Japan
14	0.44	**0.90**	0.19	0.25	0.42	**0.57**	0.74	0.61	0.44
13	0.41	0.80	0.23	0.26	0.55	0.19	0.85	0.68	0.47
12	0.53	0.89	0.30	0.26	0.29	0.44	**0.91**	0.72	0.60
11	0.44	0.82	0.52	0.28	0.56	0.28	0.89	0.75	0.38
10	0.49	0.82	0.46	0.27	0.31	0.26	0.79	0.80	**0.72**
9	0.40	0.79	0.48	0.31	0.38	0.27	0.77	0.85	0.37
8	0.51	0.76	0.51	0.31	0.52	0.32	0.79	0.88	0.61
7	**0.53**	0.76	0.51	0.34	0.25	0.40	0.85	0.92	0.47
6	0.38	0.75	0.55	0.36	0.51	**0.51**	0.86	**0.94**	0.61
5	0.36	0.68	0.45	0.41	0.31	0.28	0.81	0.94	0.47
4	0.30	0.62	0.40	0.47	0.57	0.48	0.71	0.93	0.61
3	0.25	0.63	0.47	0.43	0.33	0.43	0.62	0.94	0.50
2	0.14	0.61	0.46	0.53	0.25	0.32	0.56	0.93	0.35
1	0.13	0.60	0.37	0.55	0.43	0.38	0.56	0.92	0.37
0	0.12	0.63	0.38	0.52	0.26	0.47	0.53	0.89	0.44

	Country							
n	Norway	Saudi Arabia	Singapore	Spain	Sweden	UAE	UK	USA
14	0.44	0.41	0.09	0.34	0.69	0.45	0.75	0.87
13	0.50	0.52	0.12	0.39	0.74	0.61	0.76	0.89
12	**0.62**	0.65	0.08	0.44	0.62	0.56	0.79	0.91
11	0.56	0.62	0.05	0.50	0.44	0.58	0.78	0.92
10	0.43	0.62	0.14	0.57	0.41	0.67	0.80	0.92
9	0.39	0.62	0.17	0.65	0.49	0.70	0.86	0.92
8	0.33	**0.69**	0.10	0.71	0.64	**0.73**	0.89	0.96
7	0.23	0.52	0.02	0.78	**0.80**	0.72	**0.92**	**0.97**
6	0.27	0.39	0.06	0.83	0.79	0.70	0.88	0.97
5	0.21	0.51	0.10	0.87	0.62	0.72	0.86	0.96
4	0.09	0.51	0.09	0.91	0.43	0.66	0.85	0.93
3	0.06	0.52	0.05	0.92	0.35	0.66	0.85	0.89
2	0.01	0.60	0.07	**0.94**	0.42	0.65	0.87	0.86
1	0.00	0.58	0.10	0.93	0.56	0.66	0.87	0.85
0	0.01	0.60	0.04	0.92	0.65	0.64	0.91	0.84

Figure A9. Coefficient of determination R^2 between time series of number of daily deaths at time t and number of daily cases at time $(t-n)$, where n is in days.

References

1. Gates, B. Responding to Covid-19—A once-in-a-century pandemic? *N. Engl. J. Med.* **2020**, *382*, 1677–1679. [CrossRef] [PubMed]
2. Hoehl, S.; Rabenau, H.; Berger, A.; Kortenbusch, M.; Cinatl, J.; Bojkova, D.; Behrens, P.; Böddinghaus, B.; Götsch, U.; Naujoks, F.; et al. Evidence of SARS-CoV-2 infection in returning travelers from Wuhan, China. *N. Engl. J. Med.* **2020**, *382*, 1278–1280. [CrossRef] [PubMed]
3. Wang, C.; Pan, R.; Wan, X.; Tan, Y.; Xu, L.; Ho, C.S.; Ho, R.C. Immediate psychological responses and associated factors during the initial stage of the 2019 coronavirus disease (COVID-19) epidemic among the general population in China. *Int. J. Environ. Res. Public Health* **2020**, *17*, 1729. [CrossRef] [PubMed]
4. Brzezinski, A.; Deiana, G.; Kecht, V.; Van Dijcke, D. The COVID-19 Pandemic: Government vs. Community Action across the United States. *Covid Econ. Vetted Real-Time P.* **2020**, *7*, 115–156.
5. Sterman, J.D. *Business Dynamics: Systems Thinking and Modeling for a Complex World*; The McGraw-Hill Companies: New York, NY, USA, 2000.
6. Bertone, E.; Sahin, O.; Richards, R.; Roiko, A. Extreme events, water quality and health: A participatory Bayesian risk assessment tool for managers of reservoirs. *J. Clean. Prod.* **2016**, *135*, 657–667. [CrossRef]
7. Bertone, E.; Sahin, O.; Stewart, R.A.; Zou, P.X.; Alam, M.; Hampson, K.; Blair, E. Role of financial mechanisms for accelerating the rate of water and energy efficiency retrofits in Australian public buildings: Hybrid Bayesian Network and System Dynamics modelling approach. *Appl. Energy* **2018**, *210*. [CrossRef]

8. Sahin, O.; Suprun, E.; Richards, R.; Salim, H.; MacAskill, S.; Heilgeist, S.; Stewart, R.; Rutherford, S.; Beal, C. Navigating the Wicked Complexity of the COVID-19 Pandemic. *Syst. Commun.* **2020**. accepted for publication.

9. Fenton, N.; Neil, M. *Risk Assessment and Decision Analysis with Bayesian Networks*; CRC Press: New York, NY, USA, 2013.

10. Hale, T.; Webster, S.; Petherick, A.; Phillips, T.; Kira, B. Oxford COVID-19 Government Response Tracker, Blavatnik School of Government. 2020. Available online: https://www.bsg.ox.ac.uk/research/research-projects/coronavirus-government-response-tracker (accessed on 5 June 2020).

11. Hale, T.; Petherick, A.; Kira, B.; Angrist, N.; Phillips, T. *Variation in Government Responses to COVID-19*; BSG Working Paper Series; Blavatnik School of Government: Oxford, UK, 2020.

12. Wilson, N.; Kvalsvig, A.; Barnard, L.T.; Baker, M. Estimating the Case Fatality Risk of COVID-19 using Cases from Outside China. *medRxiv* **2020**. [CrossRef]

13. Bendavid, E.; Mulaney, B.; Sood, N.; Shah, S.; Ling, E.; Bromley-Dulfano, R.; Lai, C.; Weissberg, Z.; Saavedra, R.; Tedrow, J.; et al. COVID-19 Antibody Seroprevalence in Santa Clara County, California. *medRxiv* **2020**. [CrossRef]

14. Volpato, S.; Landi, F.; Incalzi, R.A. Frail Health Care System for an Old Population: Lesson form the COVID-19 Outbreak in Italy. *J. Gerontol. Ser. A* **2020**. [CrossRef] [PubMed]

15. Pueyo, T. Coronavirus: The Hammer and the Dance. Available online: https://medium.com/@tomaspueyo/coronavirus-the-hammer-and-the-dance-be9337092b56 (accessed on 5 May 2020).

16. Hensvik, L.; Skans, O. *COVID-19 Crisis Response Monitoring*; IZA Institute of Labor Economics: Bonn, Germany, 2020.

17. Wang, J.; Tang, K.; Feng, K.; Lv, W. High temperature and high humidity reduce the transmission of COVID-19. *SSRN* **2020**. [CrossRef]

18. Rocklöv, J.; Sjödin, H. High population densities catalyse the spread of COVID-19. *J. Travel Med.* **2020**, *27*. [CrossRef] [PubMed]

19. YouGov. YouGov International COVID-19 Tracker. 2020. Available online: https://yougov.co.uk/topics/international/articles-reports/2020/03/17/YouGov-international-COVID-19-tracker (accessed on 5 June 2020).

20. Acaps. #COVID19 Government Measures Dataset. 2020. Available online: https://www.acaps.org/covid19-government-measures-dataset (accessed on 5 June 2020).

21. Bayes, M.; Price, M. An essay towards solving a problem in the doctrine of chances. By the late Rev. Mr. Bayes, F.R.S. communicated by Mr. Price, in a letter to John Canton, A.M.F.R.S. *Philos. Trans.* **1763**, *53*, 370–418.

22. Chen, S.; Pollino, C. Good practice in Bayesian network modelling. *Environ. Model. Softw.* **2012**, *37*, 134–145. [CrossRef]

23. Uusitalo, L. Advantages and challenges of Bayesian networks in environmental modelling. *Ecol. Model.* **2007**, *203*, 312–318. [CrossRef]

24. Langarizadeh, M.; Moghbeli, F. Applying naive bayesian networks to disease prediction: A systematic review. *Acta Inform. Med.* **2016**, *24*, 364. [CrossRef] [PubMed]

25. Solares, C.; Sanz, A.M. Different Bayesian network models in the classification of remote sensing images. In Proceedings of the Intelligent Data Engineering and Automated Learning-IDEAL 2007, Birmingham, UK, 16–19 December 2007.

26. Friedman, N.; Geiger, D.; Goldszmidt, M. Bayesian network classifiers. *Mach. Learn.* **1997**, *29*, 131–163. [CrossRef]

27. Norsys. Netica APIs (Application Programmer Interfaces). Available online: https://www.norsys.com/netica_api.html (accessed on 6 March 2020).

28. Neapolitan, R. *Probabilistic Reasoning in Expert Systems: Theory and Algorithms*; John Wiley: New York, NY, USA, 1990.

Article

Cognitive Network Science Reconstructs How Experts, News Outlets and Social Media Perceived the COVID-19 Pandemic

Massimo Stella [1,2]

[1] Department of Computer Science, University of Exeter, Exeter EX4 4QF, UK; massimo.stella@inbox.com
[2] Complex Science Consulting, Via Amilcare Foscarini 2, 73100 Lecce, Italy

Received: 27 August 2020; Accepted: 26 October 2020; Published: 29 October 2020

Abstract: This work uses cognitive network science to reconstruct how experts, influential news outlets and social media perceived and reported the news "COVID-19 is a pandemic". In an exploratory corpus of 1 public speech, 10 influential news media articles on the same news and 37,500 trending tweets, the same pandemic declaration elicited a wide spectrum of perceptions retrieved by automatic language processing. While the WHO adopted a narrative strategy of mitigating the pandemic by raising public concern, some news media promoted fear for economic repercussions, while others channelled trust in contagion containment through semantic associations with science. In Italy, the first country to adopt a nationwide lockdown, social discourse perceived the pandemic with anger and fear, emotions of grief elaboration, but also with trust, a useful mechanism for coping with threats. Whereas news mostly elicited individual emotions, social media promoted much richer perceptions, where negative and positive emotional states coexisted, and where trust mainly originated from politics-related jargon rather than from science. This indicates that social media linked the pandemics to institutions and their intervention policies. Since both trust and fear strongly influence people's risk-averse behaviour and mental/physical wellbeing, identifying evidence for these emotions is key under a global health crisis. Cognitive network science opens the way to unveiling the emotional framings of massively read news in automatic ways, with relevance for better understanding how information was framed and perceived by large audiences.

Keywords: COVID-19; computational cognitive science; semantic networks; text mining; social media mining; emotions

---

1. Introduction

On 11 March 2020 the World Health Organization (WHO) declared COVID-19 a pandemic. Simultaneously, a secondary infodemic of COVID-19 news flooded information systems, overwhelming large audiences with a deluge of content about the COVID-19 spread [1]. In addition to this infodemic of fake news, even fact-checked articles reported a wide variety of contrasting views about the pandemic, e.g., diminishing the threat posed by the pathogen, creating alarm about the impact of the novel coronavirus over the economy or trying to convey the importance of social distancing for containing virus spreading. This massive proliferation of multiple views about the novel coronavirus overwhelmed large audiences with limited attention, ultimately promoting anxiety and inhibiting people from understanding the exact and correct dynamics of the pandemic [2].

In order to reduce anxiety and stress, identifying the contents promoted by this "battle of ideas" is urgently needed, also given how semantic and emotional content can deeply influence how we perceive and respond to massive events [2–4].

This manuscript outlines how recent tools from computational cognitive science [5,6] can crucially reconstruct how experts, news media and social media perceived and discussed the COVID-19 pandemic. Theoretical grounding is provided by semantic frame theory, which indicates that the meaning attributed in language to individual words or concepts is fundamentally built over conceptual associations with other concepts [7–9]. Hence, meaning cannot be attributed to words/concepts in isolation but it rather requires access to their semantic frame, i.e., the set of concepts linked to a given conceptual entity as occurring in language. Therefore, understanding the cognitive reflection attributed by individuals to the real world in terms of perceived meaning depends on the semantic frame in which concepts are entwined together. For instance, in a text if "pandemic" is conceptually related to other jargon like "disease", "spreading" and "pathogen" then the text author(s) framed "pandemics" as a phenomenon related to disease diffusion. Conceptual relationships might be of different types, like word co-occurrences (see also the interesting historical investigation of "risk" in [10]) or syntactic/semantic links (like words specifying each other in a sentence or possessing the same meaning, see also [5]). Semantic frames arise from the associative structure of language and contain both semantic and emotional content [9,11]. Recent research pointed out how different populations can reconstruct semantic frames containing very different emotions around the very same concept, e.g., online users related the very same hashtag #coronavirus to more pronounced fear-eliciting content when discussing the medical emergency and to more trustful concepts when discussing lockdown and other measures preventing contagion [12]. The identification of the emotional content of a given portion of language is called emotional profiling and it has been recently applied also to the investigation of online perceptions of COVID-19 [12–15].

This manuscript combines semantic frame theory with emotional profiling in order to reconstruct the plurality of views and emotions on the COVID-19 pandemic that were presented to millions of individuals by: (i) the authoritative WHO declaration of 11 March 2020, (ii) influential news media reporting such declaration and (iii) online social media. Attention is given to the first news that officially declared COVID-19 as a global pandemic for the whole world and not limited to Wuhan or China only. This news, "COVID-19 is a pandemic", started with the declaration by the WHO dated 11 March 2020 and it then reverberated across the above information channels. The structure of the paper includes a Methods section, outlining the linguistic datasets and the network analysis used here, followed by a Results section, reporting on the semantic frames and emotional profiles across eight basic emotions [11] of all the above declarations/articles/social media discourse. A Discussion section comments on the plurality of semantic views found in this exploration. The Discussion also interprets such results in light of relevant theories from cognitive science and other studies about the psychological impact of COVID-19 in relation to anxiety management, grief elaboration and psychological distress.

2. Material and Methods

This exploratory study focused on how COVID-19 was perceived by different information channels and investigated a corpus including three different sources of textual data: 1 speech transcript (WHO declaration), 10 news media articles on the same news (i.e., reporting the WHO declaration) and 37,500 tweets from social media (discussing the COVID-19 pandemic in the immediate aftermath of the WHO declaration). Notice that the requirement for articles to be on the same news crucially reduced the sample size of investigated texts.

This section outlines relevant information about the datasets and methods used in the main text.

2.1. Data Access: SPEECH Transcript and News Media Articles

The main analysis investigated the structure of linguistic knowledge as expressed in texts. This study did not generate any new data. The 10 news media analysed here were identified as influential by using a Google query as implemented in Mathematica 11.3. The Google query "COVID-19 WHO pandemic" on 1 April 2020 provided the top-ranked ten articles reported in Table 1, mostly produced between 10 and 12 March 2020. Notice that all investigated articles were:

(1) Produced, published and updated between 10 March and 22 March, which is the immediate aftermath of the WHO declaration of the COVID-19 outbreak being a pandemic (dated 10 March 2020). A total of 9 out of 10 articles were produced and updated between 10 March and 12 March, guaranteeing a temporal coherence of the considered dataset, i.e., a collective of articles being produced within the same time window of the investigated Twitter dataset (see also Section 2.2);

(2) As reported in their titles, all the investigated articles focused on the same news of COVID-19 being declared a pandemic by the WHO. This is an important indicator that the considered news articles did not focus on local happenings but rather analysed and interpreted the COVID-19 pandemic in light of the WHO declaration.

Table 1. News media articles investigated in this study, with online links and labels (Media 1–10).

Media ID and Link	News Outlet	Article Title	Number of Words
Media1	ABC News	What the WHO pandemic declaration means.	748
Media2	Business Insider	The coronavirus is officially a pandemic.	576
Media3	BBC	Coronavirus: What is a pandemic and why use the term now?	318
Media4	Channel News Asia	Threat of coronavirus pandemic now "very real": WHO	553
Media5	New Scientist	COVID-19: Why won't the WHO officially declare a coronavirus pandemic? (Updated 11 March)	801
Media6	National Geographic	Coronavirus is officially a pandemic.	1588
Media7	CNBC	World Health Organization declares the coronavirus outbreak a global pandemic	939
Media8	Telegraph	Coronavirus outbreak declared a pandemic: what does it mean, and does it change anything?	1387
Media9	Times	World Health Organization Declares COVID-19 a "Pandemic."	625
Media10	Washington Post	WHO declares a pandemic of coronavirus disease COVID-19	1035

The speech transcript of the WHO declaration of COVID-19 being a pandemic was obtained from the official website of the WHO (www.who.int, Last Accessed: 22 September 2020) and it included 778 words.

2.2. Data Access: Social Media Tweets

This analysis featured 37,500 Italian trending tweets gathered by Complex Science Consulting through the Twitter-approved account @ConsultComplex between 11 March and 17 March 2020. All the investigated tweets had to feature the hashtag #pandemic together with one of the following hashtags: #coronavirus, #COVID-19, #COVID. For instance, tweets featuring both #pandemic and #COVID were included in the analysis. Only tweets identified as trending by the Mathematica 11.3 "Popular" flag in the Twitter crawler ServiceConnect [] were considered in this analysis. The analysis focused on Italy given its resonance across the media due to the dramatic escalation of COVID-19 contagions in March 2020. A different dataset was also crawled by Complex Science Consulting within the same time period and investigated separately by [12].

Media links and pictures were discarded. Whereas in other studies, like [12], hashtags were used in order to infer different topics embedded in tweets, in which hashtags were integrated with other words in text. Hashtags were translated in Italian words, where possible, by using a simple overlap rule between the hashtag content with no \# symbol and Italian words (e.g., #COVID became "COVID"). Tweets in languages different from Italian were not considered in this analysis. The resulting dataset included 37,500 tweets, each one treated as a string of text. No user-level information was used in the investigation (e.g., usernames, number of followers, etc.). No tweet-level information was used (e.g., number or likes, number of retweets, etc.), except for the timestamp, which was used in order to select only messages produced between 11 March and 17 March 2020.

Although this work focused on Italian social media, the investigated tweets were influenced by the WHO declaration too and they were gathered in the next few days after the declaration itself. Hence, the social discourse analysed here cannot be considered independent from the views portrayed in news articles. Establishing a direct causal link between exposure to news articles and social media reactions goes beyond the scope of this explorative study and it represents a delicate research question that started being successfully explored only in recent studies [16,17]. Instead, this investigation focuses on comparing the semantic frames and emotional perceptions about the COVID-19 pandemic in the aftermath of the WHO declaration across different online sources.

2.3. Forma Mentis Networks as Knowledge Graphs Extracted from Text

This manuscript adopted the recent framework of forma mentis networks—as already introduced in previous studies—that used automatic text processing for reconstructing how social media discussed the gender gap in science [5] and for exploring online perceptions and emotions in Italy after the release of national lockdown [13]. On the one hand, [5] also showed that textual forma mentis networks in annotated texts are successful in determining the topic of a text. On the other hand, [13] showed that textual forma mentis networks are sensitive enough to highlight flickering emotions over time in the social discourse around specific topics and hashtags. The interested reader is referred to these prior works for more details. Here, the methodology of textual forma mentis networks is reported in a concise yet self-contained manner. Based on cognitive data and text processing, forma mentis networks are knowledge graphs [5] representing the mental lexicon, a cognitive system storing linguistic information and driving word acquisition and use [6,9]. Representing conceptual associations between words as a network of nodes (words) and links (syntactic/semantic associations), textual forma mentis networks (TFMNs) extract both knowledge and affective patterns, as embedded in the text and are representative of a given author's mindset. Other approaches also represented texts as complex networks though they focused mainly on co-occurrences between words, did not consider the cognitive/emotional dimensions of words but were successful in identifying patterns like writing styles [18]. Several successful approaches also used syntactic relationships in order to investigate how words tend be assembled in sentences in order to minimise dependency crossings and thus, potential confusions about understanding knowledge and its meaning (for a review, see [19]). Here, TFMNs focus on associative knowledge, linking together concepts according to meaning specifications provided in language (see [5]). When we read sentences, we do not explicitly observe such a conceptual network, but we mentally reconstruct it in order to understand meaning. Consider the example "pandemics and diseases are terrible", a sentence specifying that both the concepts "pandemic" and "disease" have a specific feature, i.e., being "terrible". A textual forma mentis network would represent such a sentence as a network connecting the conceptual nodes "pandemic" and "disease" to "terrible". In other words, textual forma mentis networks highlight the syntactic network of conceptual associations underlying a given text and conveying a given meaning. TFMNs thus reconstruct the syntactic and semantic structure of texts, also providing the advantage of visualising, highlighting and exploring conceptual associations in texts under the lens of network science. Syntactic relationships were extracted by using the language models available in TextStructure[], a command implemented in Mathematica 11.3. For instance, in the sentence "hospitals are full" there is a syntactic relationship specifying a property (being "full") of "hospitals", thus linking "hospital" (stem for "hospitals") and "full". Semantic relationships indicating meaning overlap (i.e., "disease" and "sickness" being synonyms in some contexts) were retrieved by using WordNet 3.0 in English and its Italian translation [20]. All words in texts were stemmed in order to represent different inflected forms with the same concept (e.g., "talking", "talked" and "talks" all representing "talk"). English words were stemmed by using the WordStem[] function implemented in Mathematica 11.3. Stemming considers only the core of a word, discarding inflections and suffixes (e.g., "learning" and "learns" would be represented with the same stem "learn"). In Italian, words were stemmed by using the Snowball C stemming algorithm as implemented in R. Italian words were translated into English for easier visualisation.

Figure 1 provides a visual scheme of the different operations leading from textual data to reconstructed, quantitative perceptions of a given topic or keyword. It must be underlined that the application of the same unsupervised methodology across different texts enables a quantitative comparison of the semantic content and emotional portrayals included in the language used by different authors. The cognitive datasets enabling emotional profiling are described in the next Section.

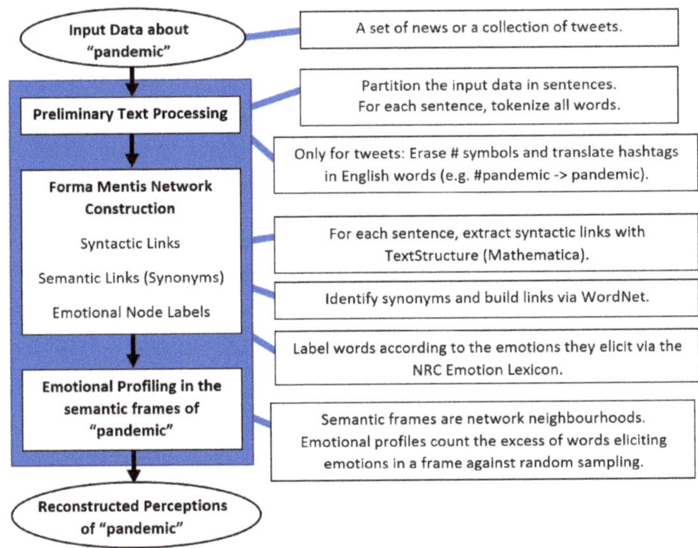

Figure 1. A scheme of the different phases leading to reconstructing emotional profiles and semantic content in texts via textual forma mentis networks.

2.4. Emotional Profiling and Cognitive Datasets

Combined with the Emotion Lexicon by Mohammad and Turney [21], TFMNs can identify the emotional profile of a given concept in a collection of texts. The emotional profile of a concept C, here was defined as the number of words eliciting a given emotion associated/linked to C, i.e., words in the network neighbourhood of C [5,12,13]. Hence, the emotional profile of a concept in a forma mentis network was computed by counting how many words elicited a given emotion in the network neighbourhood of a concept (e.g., 35% of the neighbours of "pandemic" elicited fear). The emotions considered here were the ones included in the NRC Emotion Lexicon, namely: anger, anticipation, disgust, fear, joy, sadness, surprise and trust. In counting the associations giving rise to a given emotional profile, negations were considered by associating antonyms to negated words. In presence of words exclusively linked to a negation (e.g., "love" and "not"), antonyms directly substituted negated words (e.g., "hate" was considered for the emotional profiling, replacing "love" and "not"). In the presence of words linked to a negation and to other concepts of the same sentiment (e.g., "love" connected to both "not" and "faith"), both antonyms and negated words were considered (e.g., "hate" and "love" were both considered for the emotional profiling, replacing "love" and "not"). The NRC Emotion Lexicon in Italian was obtained from the English one by adopting a consensus translation with services including Google Translate, Bing Translator and DeepL. Where at least two out of three translator services provided the same translation for a word, that one was used. In all other cases translation was human coded by the author. On a random sample of 200 words, the accuracy of translation was estimated to be 93%. This required also an additional human coding of the translation, checking for and correcting mistakes according to the author's expertise.

Counting words provides an emotional profile indicating how rich a given semantic frame/network neighbourhood is in terms of words eliciting the above emotions. However, the NRC lexicon features more words eliciting certain emotions and fewer words eliciting other emotional states, so that considering word counting by itself might be provide biased results if no null model is considered. In the current study, the null model for emotional profiles was random word sampling from the NRC lexicon, i.e., sampling at random from the lexicon as many words as observed in the emotional profile and computing the relative random emotional profile. Z-scores were computed between the empirical emotional profiles of "pandemic" and a distribution of 1000 random samples. The confidence interval relative to a 0.05 significance level was adopted as the neutrality range, i.e., a range where no emotional pattern was identified.

Notice that emotional profiles were visualised through emotion wheels, following the visualisation method originally proposed by Plutchik [11]. A blue/grey/red colour scheme was used in order to cluster together negative emotions (in red), neutral emotions (in grey) and positive emotions (in blue).

3. Results

Semantic frames and emotions, as attributed in the processed texts, highlight interesting patterns on how the COVID-19 pandemic was announced by WHO and discussed by news outlets and social media users.

3.1. Investigating the Knowledge and Emotions in the Whole WHO Declaration

Figure 2 visualises the WHO declaration of "COVID-19 is a pandemic" in terms of a knowledge graph, a textual forma mentis network representing the mindset (in Latin forma mentis) portrayed in the declaration in terms of concepts (nodes) linked by semantic and syntactic relationships (cf. Methods). Larger nodes are relative to concepts with a higher closeness centrality (see [22]), i.e., connected by fewer links to all other concepts in the network. Since previous studies indicated how closeness in forma mentis networks captures semantic prominence (see [5]), the knowledge graph indicates how the WHO declaration mainly revolved around "country". Other prominent concepts (i.e., high in closeness centrality) were "we", "not", "take", "case", "change", "tell" and "numbers", indicating that the speech adopted a first-person narrative revolving around countries and the change (of the pandemic) due to the registered numbers of cases. This is not surprising but comes as a confirmation that a network metric (closeness centrality) can identify key aspects of an apparently unstructured piece of text (in this case the WHO declaration of COVID-19 being a pandemic).

In addition to this reality check, the knowledge graph also visually represents key aspects of a text. For instance, in Figure 2 words are clustered in communities as identified by a Louvain algorithm [22]. Communities are defined as network regions made of tightly interconnected nodes and in a forma mentis network a community can provide contextual knowledge describing the meaning attributed to concepts across sentences [5]. By considering the community including "pandemic" one can notice that the WHO speech presented the concept of a pandemic to the general audience in relationship to "health", "change", "crisis" and even "economy": the WHO presented the pandemic as a challenging crisis not only for global health but also for the economy. Positive terms like "protect", "together" and "calm" appeared within the same community of "pandemic" and indicate a tendency for the speech to call for calmness and a communal protection against such crisis.

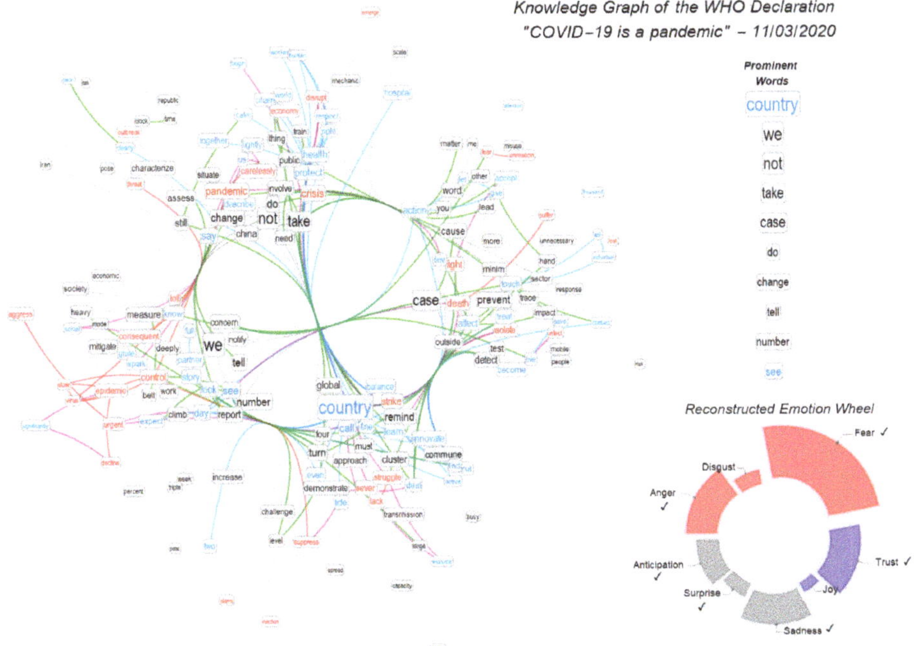

Figure 2. Left: Knowledge graph representing syntactic associations between words of positive (cyan), negative (red) and neutral (black) sentiment and semantic overlap between synsets (green) as extracted from the WHO Declaration of COVID-19 being a pandemic, 11 March 2020. Larger font size indicates higher closeness centrality and conceptual prominence in the text (see Section 2). Top right: Prominent words in the discourse as detected from syntactic and semantic associations. Bottom right: Emotion wheel, counting how many words elicited a given emotion in the knowledge graph. Emotions stronger than random expectation were marked with a check (Z-test against random word sampling, significance level of 0.05).

The overall emotional profile of the speech is reported in Figure 2 (bottom, right). The words adopted in the WHO declaration elicited strong emotional patterns of fear, anger and sadness but also levels of trust higher than random expectation. Combinations of fear and sadness indicate concern, according to Plutchik's wheel of emotions [11], but what about the combination of positive and negative emotions? The co-existence of concern and trust indicates a clear narrative strategy pursued by the WHO in their speech: to clearly explain the dangers posed by the COVID-19 emergency to the general public, while subsequently adopting also trustful language focusing on how to face the crisis itself.

3.2. Investigating Knowledge and Emotions around "Pandemic" across Texts

Forma mentis networks were built not only for the WHO declaration, but for all the considered articles and aggregated tweets. Rather than investigating the whole networks, attention was given to the individual semantic frames, i.e., the network neighbourhoods, of "pandemic" across texts. Figure 3 (top) reports the semantic frames and emotional profiles of "pandemic" in the WHO declaration, within two of the 10 investigated articles and within social discourse. Figure 3 (bottom) provides the complete spectra of emotional profiles across all the investigated media.

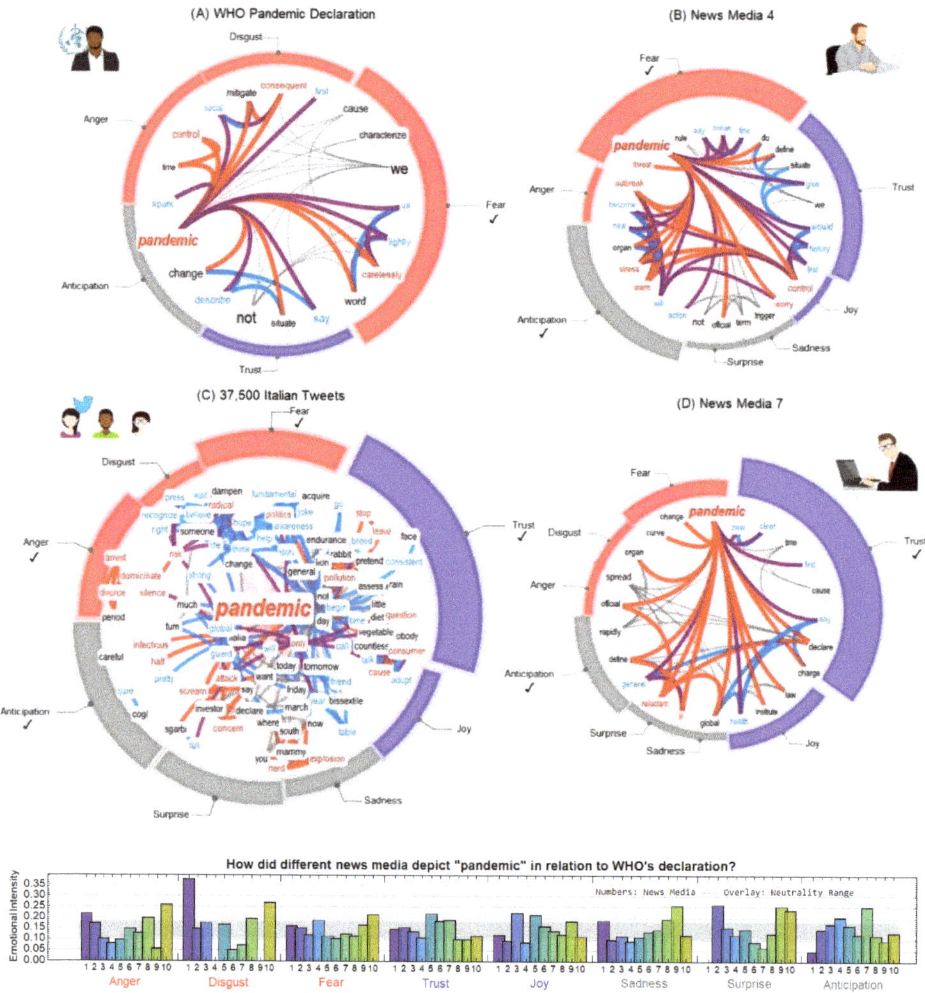

Figure 3. Top: Syntactic associations between positive (cyan), negative (red) and neutral (black) words as used in the WHO pandemic declaration (**A**), two influential news media reporting the WHO's declaration (**B,D**) and 37,500 tweets with #pandemic posted right after the WHO's declaration (**C**). Emotions stronger than random expectation were marked with a check (z-test against random word sampling, significance level of 0.05). Larger concepts were more prominent in the semantic structure of the used language. Bottom: Emotional intensities of 10 different news media are reported. The grey bar in the foreground represents the error bound of neutral intensities (see Section 2).

As reported in Figure 3, the WHO declaration pictured "pandemic" as a phenomenon linked to control and mitigation, with associated concepts eliciting a moderate level of fear and expressing concern about the COVID-19 pandemic. Hence, the overall concern found in the WHO's whole declaration is present also around the specific semantic framing attributed by the WHO to "pandemic", as expected from a speech centred around massive pathogen diffusion.

The other semantic frames and emotional profiles reported here are the most dissimilar from the WHO's one. News media four provided a highly polarised perception of "pandemic", combining fear-inducing links to "threat" with more trustful concepts about the future, a combination indicating submission, apprehension and passive acceptance of COVID-19 being a pandemic. Instead, news media seven depicted a more neutral and trustful frame for the COVID-19 pandemic, using neutral jargon, explaining the epidemiological definition of a pandemic and conveying emotions of trust into waking up, getting ready and taking action. These patterns were absent in the original WHO declaration and have thus to be considered as cognitive alterations attributed to the way individual journalists framed the news "COVID-19 is a pandemic". It should be stressed that these alterations emerge from a purely quantitative modelling of texts and do not require any human coding or reading of the underlying texts.

Social media strongly debated over the COVID-19 pandemic. Here, attention was given to Italy, the first European country struck by COVID-19 and reacting with a national lockdown (see [12]). In 37,500 Italian tweets (see Section 2), "pandemic" spawned a social discourse featuring contrasting, affect-polarising concepts like "help", and "hope" together with "risk" and "attack". The resulting emotional profile was strongly trustful, dominated by a positive awareness of the coronavirus and the need to offer assistance against it, but also mixed with patterns of fear (against the risks of the pandemics), anger (against a global threat) and anticipation (for a better future).

Anger and fear are both natural evolutionary mechanisms for coping with threats [11] and are also natural responses in elaborating grief about a negative event, as indicated by the Kübler-Ross model [23]. These emotions are therefore expected when perceiving a pandemic and were also strongly detected in other studies focusing on different datasets related to social discourse about the COVID-19 pandemic (see [15]). Interestingly, in here grief-related emotions attributed to "pandemic" were found to be dominated by more positive ones, expressing awareness and willingness to act upon the recently declared health emergency. In this way, the perception of the "pandemic" itself is quantitatively found to be way richer in positive/trustful concepts than the framing portrayed by the WHO.

4. Discussion and Conclusions

The same event can be described in text through different semantic frames, eliciting different emotions and perceptions. The real challenge in AI is quantifying and reconstructing these frames without human coding and in the absence of training data. This manuscript tested knowledge modelling via cognitive network science, as a quantitative way for automatically identifying key concepts and their emotional perceptions in texts, e.g., articles and social media messages. The semantic and emotional frames reconstructed here around the news "COVID-19 is a pandemic" open quantitative ways for tracking how COVID-19 was discussed by experts, news media and social networks.

Fear, anger and trust about the pandemic were detected in news outlets and social media reaching large audiences. It is crucially important to keep track of any massive release of information vehiculating these emotions, since the latter can strongly influence an individuals' perception, mental well-being and real-world behaviour [11,24–26]. As recent studies from computational social science pointed out [27], both negative and positive emotions can give rise to emotional contagion over social media, a phenomenon where emotions pass from user to user by the mere act of reading a post/message portraying a certain emotional profile and without any real-world experience giving grounding to the transmitted emotions. Hence, emotional contagion over social media can infect unaware audiences and potentially de-stabilise public opinion about the COVID-19 crisis [1]. A recent investigation on the reopening after lockdown found that online users on Twitter tended to retweet more those messages richer in fear-eliciting jargon [13], further indicating the possibility for fear-eliciting content to give rise to emotional contagions of fear over social media.

According to cognitive science, contagions of fear and anger would have different repercussions. Responding to threats with anger greatly inhibits risk perception [24], leading to resisting risk-reduction policies (e.g., lockdown or social distancing). Instead, fear heightens risk perception, thus leading to more risk-averse behaviour [24]. However, prolonged states of fear are detrimental for correctly identifying threats and can also induce higher levels of cortisol, which impair both mental wellbeing and the immune system [26]. Hence, a deluge of fear-inspiring semantic frames, like the ones found in some news media in this exploratory study, might have concrete repercussions over the COVID-19 contagion, influencing people's immune systems and bolstering their susceptibility to pathogens. The exact quantification of this relationship between COVID-19 fear levels and health susceptibility remains an open question for future research. However, it should be noted that fear can also assume a positive role in hampering the pandemics. In a sample of 324 people, Harper and colleagues [28] recently found that moderate levels of fear were functional in promoting public health compliance and boosting positive behavioural changes (e.g., social distancing) aimed at containing the contagion. This positive role played by fear aligns with the moderate levels of fear found in the semantic frame of "pandemic" in the WHO declaration, providing further support to the narrative expedient adopted by the WHO of raising concern about the novel coronavirus, with concerning jargon related to epidemics and also including mitigation measures.

Trust was identified in the semantic frames of "pandemic" as discussed by social media and some news outlets. The reconstructed semantic frame reported here indicated how news outlets channelled trust through scientific knowledge, a "scientific explanation" strategy useful for improving individuals' awareness about the mechanisms driving the pandemic and its containment. Recent studies also outlined how eliciting trust is a good strategy not only for reducing the incidence of fear and anger but also in order to boost compliance with regulations and responsible health behaviour [29]. The "scientific explanation" strategy was not found in the semantic frames extracted from social media discourse, where trust-eliciting concepts were not linked to scientific jargon but were rather connected to "politics". This confirms that trust can indeed come from different sources, as also channelling feelings of trust towards politics can make large audiences comply with public health orders like respecting social distancing [29,30]. Notice that higher levels of trust towards science and politics, as outlined here with this study, were also found by previous investigations targeting populations under recent lockdown caused by the COVID-19 pandemic [12,30].

Tracking the levels of perceived trust around the COVID-19 pandemic and identifying the sources of such trust, be it scientific knowledge or politics, is a crucial challenge for better understanding compliance with regulations. In fact, audiences with little trust in the institutions can mine the credibility of governments, both fostering divisive politics and hampering nationwide adoption of strategies for pandemic containment [29].

It should be noted that trust and fear are not direct opposites. According to the Atlas of Emotions [11], trust and awareness can turn anger into a constructive motivation towards facing the threat or source of anger itself—in this case the COVID-19 pandemic. This interconnectedness suggests a highly nuanced perception of social media users towards facing the COVID-19 pandemic, as also found within other studies focusing on Twitter data [12,14] and other social media platforms [1,15].

The above points outline a complex landscape of real-world implications for the different emotional perceptions of the COVID-19 pandemic. The current semantic/emotional frames can be a valid compass for navigating such landscape, with some limitations. This exploratory study considers only textual information, thus neglecting other types of communication through visual or verbal cues. Nonetheless, the plurality of perceptions quantified here, in a statistically robust way, indicate how already processing only textual sources might provide richly complex semantic frames and emotional profiles about how the COVID-19 pandemic is perceived, discussed and reported to large audiences.

Another limitation of this study was neglecting how potential automated accounts might have biased social discourse. Recent investigations [3,31] found that social bots, i.e., user accounts piloted by automatic software, were capable of influencing human behaviour by fostering negative, conspiratorial and anxious perceptions in online platforms. Applying the same tools of cognitive network science explored and tested here to the investigation of the semantic frames produced by social bots represents an interesting future research direction.

Besides the above limitations, the main advantage of cognitive network science is its ability to capture linguistic associations, semantic frames and emotions in a transparent way, enabling a visualisation of conceptual associations in texts and without requiring training data for machine learning.

Future research should build upon cognitive network science for a better understanding of social media, possibly in synergy with other promising and successful automatic approaches to knowledge modelling [32–37]. Access to large-scale corpora of news media articles would enable prompt identification of outlets promoting distorted mindsets ("COVID-19 is just a flu") or panic-inducing misinformation. Additional cross-validation through careful human coding should be pursued as well, with the aim of further characterising how the semantic information highlighted by textual forma mentis networks is understood and processed by individuals. This calls for future research endeavours beyond the scope of the current preliminary study but always within the growing area of cognitive data science.

A richer understanding of the context of promoting negative perceptions should also be pursued in the analysis of political debate, like in [33], where cognitive networks can unveil how political parties or politicians discuss the COVID-19 emergency. The current methodology can also identify how people perceive health risks and concrete symptoms of COVID-19, integrating recent powerful analyses extracting the symptoms of the novel coronavirus from social discourse [35] and complementing interesting dynamical patterns of grief elaboration and COVID-19 recently unearthed in social platforms by Aiello and colleagues [15]. Notice that the representation of textual knowledge in tweets produced here could also be integrated with other representations based on word embedding models [32,34,36] and powering recent natural language approaches to identifying topics in tweets discussing COVID-19 [36,37]. Forma mentis networks and other models of natural language processing all aim towards the common direction of monitoring and understanding large volumes of messages with the ease of lightweight and automatic knowledge extraction methodologies.

In this time of uncertainty, about the evolution of the coronavirus pandemic and the impact of lockdowns, having a clear, misinformation-free perception of the COVID-19 pandemic is fundamental [38].The cognitive network science framework outlined in this work can tackle this challenge, effectively capturing and reproducing a plurality of views promoted about the COVID-19 pandemic. Access to such information opens new ways for policy makers to act based on how large audiences perceive the COVID-19 pandemic and its repercussions and thus represents an impactful direction for future research.

Funding: The author received no financial support for the research, authorship, and/or publication of this article.

Conflicts of Interest: The author declares no conflict of interest.

References

1. Cinelli, M.; Quattrociocchi, W.; Galeazzi, A.; Valensise, C.M.; Brugnoli, E.; Schmidt, A.L.; Zola, P.; Zollo, F.; Scala, A. The covid-19 social media infodemic. *Sci. Rep.* **2020**, *10*, 16598. [PubMed]
2. Garfin, D.R.; Silver, R.C.; Holman, E.A. The novel coronavirus (COVID-2019) outbreak: Amplification of public health consequences by media exposure. *Health Psychol.* **2020**, *39*, 355–357. [CrossRef] [PubMed]
3. Stella, M.; Ferrara, E.; De Domenico, M. Bots increase exposure to negative and inflammatory content in online social systems. *Proc. Natl. Acad. Sci. USA* **2018**, *115*, 12435–12440.
4. Gallotti, R.; Valle, F.; Castaldo, N.; Sacco, P.; De Domenico, M. Assessing the risks of "infodemics" in response to COVID-19 epidemics. *arXiv* **2020**, arXiv:2004.03997.
5. Stella, M. Text-mining forma mentis networks reconstruct public perception of the STEM gender gap in social media. *PeerJ Comput. Sci.* **2020**, *6*, e295. [CrossRef]
6. Siew, C.S.; Wulff, D.U.; Beckage, N.M.; Kenett, Y.N. Cognitive Network Science: A Review of Research on Cognition through the Lens of Network Representations, Processes, and Dynamics. *Complexity* **2019**, *2019*, 2108423. [CrossRef]
7. De Deyne, S.; Kenett, Y.N.; Anaki, D.; Faust, M.; Navarro, D. Large-Scale Network Representations of Semantics in the Mental Lexicon. In *Big Data in Cognitive Science*; Routledge/Taylor & Francis Group: Abingdon, UK, 2017; pp. 174–202.
8. Fillmore, C.J.; Baker, C.F. Frame semantics for text understanding. In Proceedings of the WordNet and Other Lexical Resources Workshop NAACL, Pittsburgh, PA, USA, 3–4 June 2001; Volume 6.

9. Vitevitch, M.S. (Ed.) *Network Science in Cognitive Psychology*; Routledge: Abingdon, UK, 2019.

10. Li, Y.; Hills, T.; Hertwig, R. A brief history of risk. *Cognition* **2020**, *203*, 104344.

11. Plutchik, R. The nature of emotions. *Am. Sci.* **2001**, *89*, 344–350.

12. Stella, M.; Restocchi, V.; De Deyne, S. #lockdown: Network-Enhanced Emotional Profiling in the Time of COVID-19. *Big Data Cogn. Comput.* **2020**, *4*, 14.

13. Stella, M. Social discourse and reopening after COVID-19: A post-lockdown analysis of flickering emotions and trending stances in Italy. *PsyarXiv* **2020**. [CrossRef]

14. Lwin, M.O.; Lu, J.; Sheldenkar, A.; Schulz, P.J.; Shin, W.; Gupta, R.; Yang, Y. Global sentiments surrounding the COVID-19 pandemic on Twitter: Analysis of Twitter trends. *JMIR Public Health Surveill.* **2020**, *6*, e19447.

15. Aiello, L.M.; Quercia, D.; Zhou, K.; Constantinides, M.; Šćepanović, S.; Joglekar, S. How Epidemic Psychology Works on Social Media: Evolution of responses to the COVID-19 pandemic. *arXiv* **2020**, arXiv:2007.13169.

16. Schmidt, A.L.; Peruzzi, A.; Scala, A.; Cinelli, M.; Pomerantsev, P.; Applebaum, A.; Gaston, S.; Fusi, N.; Peterson, Z.; Severgnini, G.; et al. Measuring social response to different journalistic techniques on Facebook. *Humanit. Soc. Sci. Commun.* **2020**, *7*, 17. [CrossRef]

17. Gozzi, N.; Tizzani, M.; Starnini, M.; Ciulla, F.; Paolotti, D.; Panisson, A.; Perra, N. Collective response to the media coverage of COVID-19 Pandemic on Reddit and Wikipedia. *J. Med. Internet Res.* **2020**, *22*, e21597.

18. Amancio, D.R. Probing the topological properties of complex networks modeling short written texts. *PLoS ONE* **2015**, *10*, e0118394. [CrossRef]

19. Baronchelli, A.; Ferrer-I-Cancho, R.; Pastor-Satorras, R.; Chater, N.; Christiansen, M.H. Networks in cognitive science. *Trends Cogn. Sci.* **2013**, *17*, 348–360. [CrossRef] [PubMed]

20. Bond, F.; Foster, R. Linking and extending an open multilingual wordnet. In Proceedings of the 51st Annual Meeting of the Association for Computational Linguistics, Sofia, Bulgaria, 4–9 August 2013; Volume 1, pp. 1352–1362.

21. Mohammad, S.M.; Turney, P.D. Emotions evoked by common words and phrases: Using mechanical turk to create an emotion lexicon. In Proceedings of the NAACL HLT 2010 Workshop on Computational Approaches to Analysis and Generation of Emotion in Text, Association for Computational Linguistics, Los Angeles, CA, USA, 5 June 2010; pp. 26–34.

22. Newman, M. *Networks*; Oxford University Press: Oxford, UK, 2018.

23. Kübler-Ross, E.; Kessler, D. *On Grief and Grieving*; Simon and Schuster: New York, USA, 2005.

24. Lerner, J.S.; Keltner, D. Fear, anger, and risk. *J. Personal. Soc. Psychol.* **2001**, *81*, 146. [CrossRef]

25. Moroń, M.; Biolik-Moroń, M. Trait emotional intelligence and emotional experiences during the COVID-19 pandemic outbreak in Poland: A daily diary study. *Personal. Individ. Differ.* **2020**, *168*, 110348.

26. Staufenbiel, S.M.; Penninx, B.W.; Spijker, A.T.; Elzinga, B.M.; van Rossum, E.F. Hair cortisol, stress exposure, and mental health in humans: A systematic review. *Psychoneuroendocrinology* **2013**, *38*, 1220–1235. [CrossRef]

27. Kramer, A.D.; Guillory, J.E.; Hancock, J.T. Experimental evidence of massive-scale emotional contagion through social networks. *Proc. Natl. Acad. Sci. USA* **2014**, *111*, 8788–8790. [CrossRef]

28. Harper, C.A.; Satchell, L.P.; Fido, D.; Latzman, R.D. Functional fear predicts public health compliance in the COVID-19 pandemic. *Int. J. Ment. Health Addict.* **2020**, 1–14. [CrossRef]

29. Khemani, S. An Opportunity to Build Legitimacy and Trust in Public Institutions in the Time of Covid-19. *World Bank Res. Policy Briefs* **2020**, 148256. [CrossRef]

30. Sibley, C.G.; Greaves, L.M.; Satherley, N.; Wilson, M.S.; Overall, N.C.; Lee, C.H.J.; Milojev, P.; Bulbulia, J.; Osborne, D.; Milfont, T.L.; et al. Effects of the COVID-19 pandemic and nationwide lockdown on trust, attitudes toward government, and well-being. *Am. Psychol.* **2020**, *75*, 618–630. [CrossRef] [PubMed]

31. Pozzana, I.; Ferrara, E. Measuring bot and human behavioral dynamics. *Front. Phys.* **2020**, *8*, 125. [CrossRef]

32. Lücking, A.; Brückner, S.; Abrami, G.; Uslu, T.; Mehler, A. Computational linguistic assessment of textbook and online learning media by means of threshold concepts in business education. *arXiv* **2020**, arXiv:2008.02096.

33. Brito, A.C.M.; Silva, F.N.; Amancio, D.R. A complex network approach to political analysis: Application to the Brazilian Chamber of Deputies. *PLoS ONE* **2020**, *15*, e0229928. [CrossRef] [PubMed]

34. Choi, M.; Aiello, L.M.; Varga, K.Z.; Quercia, D. Ten Social Dimensions of Conversations and Relationships. In Proceedings of the Web Conference 2020, Taipei, Taiwan, 20–24 April 2020; pp. 1514–1525.

35. Murray, C.; Mitchell, L.; Tuke, J.; Mackay, M. Symptom extraction from the narratives of personal experiences with COVID-19 on Reddit. *arXiv* **2020**, arXiv:2005.10454.

36. Müller, M.; Salathé, M.; Kummervold, P.E. COVID-Twitter-BERT: A Natural Language Processing Model to Analyse COVID-19 Content on Twitter. *arXiv* **2020**, arXiv:2005.07503.
37. Gencoglu, O. Large-scale, Language-agnostic Discourse Classification of Tweets During COVID-19. *arXiv* **2020**, arXiv:2008.00461.
38. Larson, H.J. Blocking information on COVID-19 can fuel the spread of misinformation. *Nature* **2020**, *580*, 306.

Publisher's Note: MDPI stays neutral with regard to jurisdictional claims in published maps and institutional affiliations.

Article

Mathematical Modeling and Simulation of the COVID-19 Pandemic

Günter Bärwolff

Department of Mathematics, Technische Universität Berlin, D-10623 Berlin, Germany;
baerwolf@math.tu-berlin.de

Received: 22 April 2020; Accepted: 6 July 2020; Published: 13 July 2020

Abstract: The current pandemic is a great challenge for several research areas. In addition to virology research, mathematical models and simulations can be a valuable contribution to the understanding of the dynamics of the pandemic and can give recommendations to physicians and politicians. Based on actual data of people infected with COVID-19 from the European Center for Disease Prevention and Control (ECDC), input parameters of mathematical models will be determined and applied. These parameters will be estimated for the UK, Italy, Spain, and Germany and used in an SIR-type model. As a basis for the model's calibration, the initial exponential growth phase of the COVID-19 pandemic in the named countries is used. Strategies for the commencing and ending of social and economic shutdown measures are discussed.

Keywords: mathematical epidemiology; SIR-type model; model parameter estimation; non-pharmaceutical intervention; dynamical systems; COVID-19/SARS-CoV2

1. Introduction

COVID-19 is a recent emerging disease caused by the emerging coronavirus. As there is no immunity to this virus, the spread of the disease has been rampant worldwide. As no serious vaccine or medication exists, it is necessary to look for effective non-pharmaceutical interventions to control the pandemic. Here, I use an SIR-type model to understand and analyze the COVID-19 pandemic with the aim of stopping or reducing the spread of the COVID-19 virus.

The dynamic development of sub-populations of susceptible (S), infected (I), and recovered (R) people in a certain region—for example, the population of a country or a part of a federation— depending on non-pharmaceutical interventions is the aim of the modeling. Deterministic models are discussed. These are simple but effective for describing the progression of the pandemic. They are able to fit the description of the average infection dynamics in macroscopic sub-populations only[1].

The main scope of this paper is the investigation of certain lockdown measures to flatten the curve of infected people over time and of the appropriate strategies for returning from lockdown to normality. To find appropriate model parameters, real data of the early stages of the pandemic are analyzed. Suggestions about favorable points in time at which to commence with lockdown measures based on the acceleration rate of infections conclude the paper.

There are also more complex deterministic models, which include sub-populations other than S, I, and R (see [1,2]), but these models have dynamic properties similar to those of the basic SIR model. On the other hand, additional data, which are not available, are needed for the extension of the basic model. This why I can perform the investigations on the basis of the SIR model without a loss of generality with respect to the aim of this paper.

---

[1]   A finer resolution of the pandemic is possible with stochastic agent-based models, which will not be discussed in this paper.

It is necessary to remark that the considered SIR model is not able to describe the full asymptotic behavior of a pandemic, as is done in [3]. In addition, the role of super-spreaders, investigated in [4] and [5], cannot be described with the basic macroscopic SIR model.

2. The Mathematical SIR Model

First, I note one important presupposition for the model. I suppose that the distribution of the included sub-populations is equal, i.e., the density is approximately constant. This is a very strict supposition, but this is acceptable, for example, for cities and congested urban areas like New York or the Ruhr area in Germany. At the beginning of the pandemic, exponential growth of the number of infected people is supposed.

In the so-called SIR model of Kermack and McKendrick [6], I denotes the infected people, S denotes the susceptible people, and R denotes the recovered people. It is a deterministic model. I constrain the investigations to the species I, S, and R only. The dynamics of infections and recoveries can be approximated by the following system of ordinary differential equations:

$$\frac{dS}{dt} = -\kappa\beta\frac{S}{N}I \tag{1}$$

$$\frac{dI}{dt} = \kappa\beta\frac{S}{N}I - \gamma I \tag{2}$$

$$\frac{dR}{dt} = \gamma I . \tag{3}$$

β represents the number of others that one infected person encounters per unit time (per day). γ is the reciprocal value of the typical time from infection to recovery. N is the total number of people involved in the epidemic disease, and $N = S + I + R$. κ is equal to one in the case of an undisturbed pandemic without any interventions or lockdowns. Later, I will specify κ as a function to describe lockdown measures.

The currently available empirical data suggest that the coronavirus infection typically lasts for some 14 days. This means that $\gamma = 1/14 \approx 0.7$.

The choice of β is more complicated and will be considered in the following.

The equation system (1)–(3) belongs to the mathematical category of dynamical systems.

3. The Estimation of β Based on Real Data

I used the European Center for Disease Prevention and Control [7] to obtain data on the people infected with COVID-19 for the period from 31 January 2019 to 8 April 2020.

At the beginning of the pandemic, the quotient S/N was nearly equal to 1. In addition, at the early stage, no one had yet recovered. Thus, I can describe the early regime using the ordinary differential equation

$$\frac{dI}{dt} = \beta I$$

with the solution

$$I(t) = I_0 \exp(\beta t) . \tag{4}$$

The logarithm of (4) leads to

$$\log I(t) = \log I_0 + \beta t .$$

Based on the table of logarithms of the infected people versus time, $(t_j, \log I_j)$, $j = 1, \ldots, k$, I look for I_0 and β which minimize the function

$$L(I_0, \beta) = \sum_{j=1}^{k} [\log I_0 + \beta t_j - \log I_j]^2 . \tag{5}$$

The solution of this linear optimization problem is trivial, and it is available in most computer algebra systems as a "black box" of the logarithmic–linear regression.

Figure 1 shows the results for the same periods as above for Spain and the UK[2].

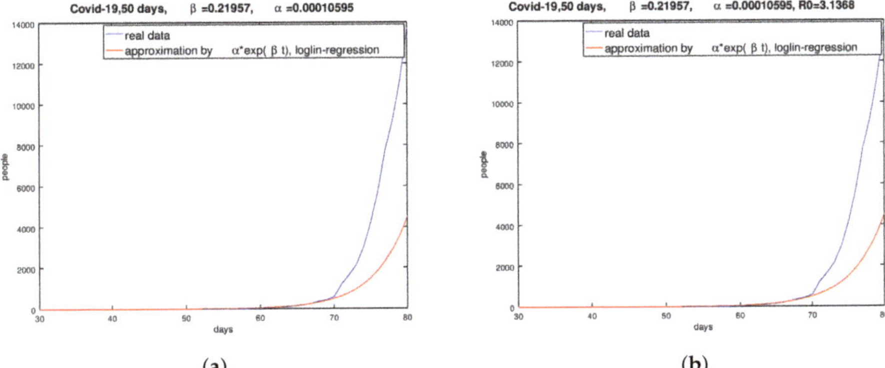

Figure 1. Course of the pandemic; β-value from the logarithmic–linear regression. (**a**) Results for Spain (31 January 2020 to 20 March 2020); (**b**) Result of the UK (30 January 2020 to 20 March 2020).

Figure 1 shows that the logarithmic–linear regression implies unsatisfactory results. It must be said that the evaluated β-values are related to the stated period. For the logarithmic–linear regression method, I guessed the respective periods for every country through a visual inspection of the graphs of the infected people versus time.

Instead of the above-used table of logarithmic values, the table (t_j, I_j), $j = 1, \ldots, k$ is used with the aim of a better approximation. I am looking for periods in the spreadsheets of infected people per day where the course can be described by a function of type (4).

Choosing a period $[t_1, t_k]$ for a certain country, I search for the minimum of the function

$$F(I_0, \beta) = \sum_{j=1}^{k} [I_0 \exp(\beta t_j) - I_j]^2 \, ,$$

i.e.,

$$\min_{(I_0, \beta) \in \mathbb{R}^2} F(I_0, \beta) \, . \tag{6}$$

I solved this non-linear minimum problem with the damped Gauss–Newton method. The results of the above-discussed logarithmic–linear method for β and α proved as good starting iterations for the Gauss–Newton method. I found the subsequent results for the considered countries. Thereby, I chose these periods for the countries with the aim of approximating the infection's progression with a good quality. Figure 2 shows the graphs and the evaluated parameters for Germany and Spain.

I found some information on the parameters of Italy in [8]—for example, $\beta = 0.25$—and I presume that this is a result of the logarithmic–linear regression by the Italian health administration.

A deeper look at the real data shows that the exponential behavior of the dynamic of the number of infected people was found only in the very beginning of the pandemic. In the German hotspot of Bavaria, I found the result $\beta = 0.22658$ for the period from 24 February to 20 April 2020 with non-linear regression. With the logarithmic–linear approach, I found a quite similar value, $\beta = 0.23$.

---

[2] The numbers of more than 4-digits at the ordinate of the following figures should be understand as numbers with a comma in the middle, for example 10,000 should be understand as 10,000 and so on.

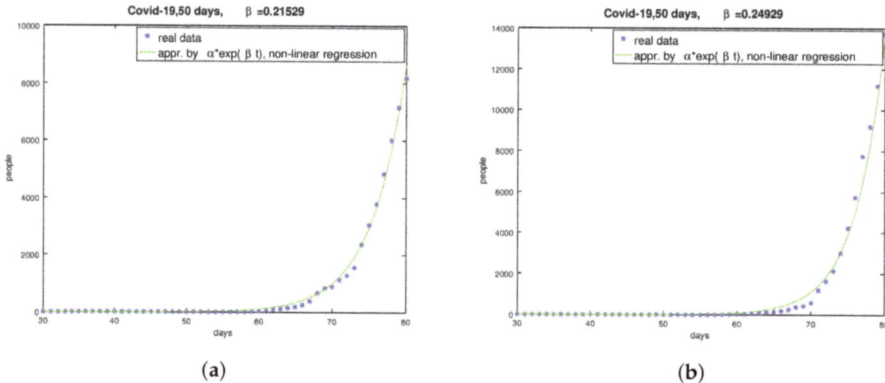

(a)

(b)

Figure 2. Results from 31 January 2020 to 20 March 2020. (**a**) German results; (**b**) Spanish results.

In conclusion, I can state that the estimation of the parameter β is complicated but successful in most of the considered countries and regions. The results of the solution of the minimum problem (6) to evaluate β are, in most cases, better than the results of the minimization of function (5) with respect to the fitting of the real data.

To illustrate the different quality and quantity of the β estimation, I use Italy as an example in Figure 3.

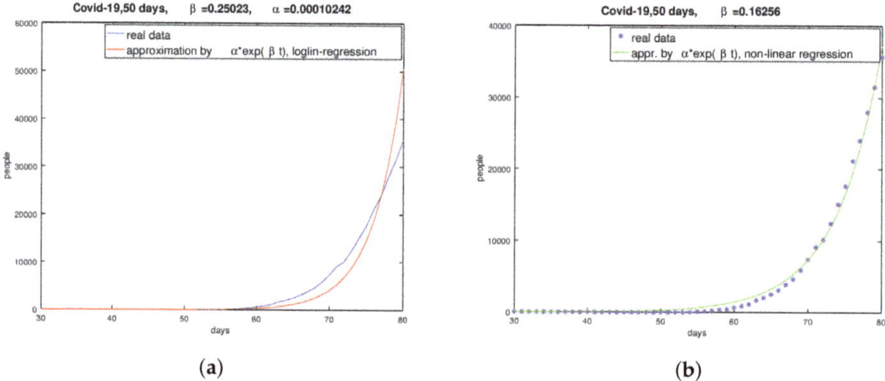

(a)

(b)

Figure 3. Italian results (from 31 January 2020 to 20 March 2020). (**a**) Results of Italy with the β-value from the logarithmic–linear regression; (**b**) Results of Italy with the β-value from the non-linear minimization.

4. Numerical Computations for Germany and Spain

I disclaim qualitative mathematical considerations like existence and uniqueness of solutions of the dynamical system of (1)–(3) and concentrate my interest on practical application and numerical experiments. The numerical solution of the ordinary differential equation system of the modified SIR model was done with a Runge–Kutta integration method of the fourth order. The independence of the time discretization of the solution method was tested by a systematic time-grid refinement. At the end, I found that time-steps of half a day could be used. For all of the following computations, the β results of the solution of the non-linear minimization problem are used.

With the choice of a β-value of 0.215 (see Figure 2a)—which is evaluated on the basis of the real data from the ECDC—and $\gamma = 0.07$, one gets the progress of the pandemic's dynamics, pictured in Figure 4a ($I0$ denotes the initial value of the I species, that is, 31 January 2020. $Imax$ stands for the maximum of I. The total number N for Germany is guessed to be 70 million). R_0 is the basis reproduction number of persons infected by the transmission of a pathogen from an infected person during the infectious time ($R_0 = \beta/\gamma$), shown in the following figures[3]. For the early 30 days, I found a β-value of 0.36 for China/Wuhan. This shows that the German situation with $\beta = 0.215$ and $R_0 \approx 3$ is moderate compared to the Chinese situation with the values $\beta \approx 0.36$ and $R_0 \approx 5$. I have to mention that these values vary compared to those found by other authors, but the relationships between the German and the Chinese values are similar.

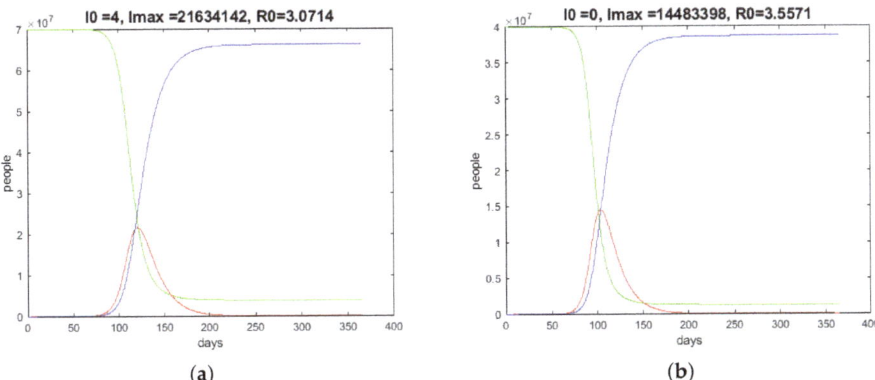

Figure 4. One-year results of Germany and Spain; S—green, I—red, R—blue. (**a**) German progression over one year, starting at the end of January 2020; (**b**) Spanish progression over one year, starting at the end of January 2020.

The data from the ECDC, the data from the German Robert Koch Institute, and the data from the Johns Hopkins University ([9]) are not really correct; thus, I have to reasonably assume that there are a number of unknown cases. It is guessed in [10] that the data cover only 15% of the real cases. Considering this, I obtained slightly changed results, and in the subsequent computations, I will include an estimated number of unknown cases in the initial values of I.

I use the β-value 0.249 (see Figure 3) and $\gamma = 0.07$ for Spain, and I get the run pictured in Figure 4. N is set to 40 million.

Let me now discuss the case of strict social distancing. To do this, I reduce the β-values after a few days to $\beta = 0.14$ for both Germany and Spain.

The results in Figure 5 compared to the results without the reduction of β (Figure 4) show the consequences. The climax of the number of infected people moved to the autumn of the year with hard inconveniences for the population, but the wanted flattening was achieved.

To investigate the influence and sensitivity of the simulation results with the parameter β and the number N (sum of infected, susceptible, and recovered people), I used the German data and a variation of these data. In Figure 6b, I see that variation of the amount N leads, more or less, to a proportional scaling[4]. The variation of β showed a non-monotone and non-linear influence of β on the results, pictured in Figure 6a.

---

[3]   The values of R_0 in all of the following figures are applied to the β-value of the beginning of the pandemic.
[4] $N = 12$ million is the population of Bavaria.

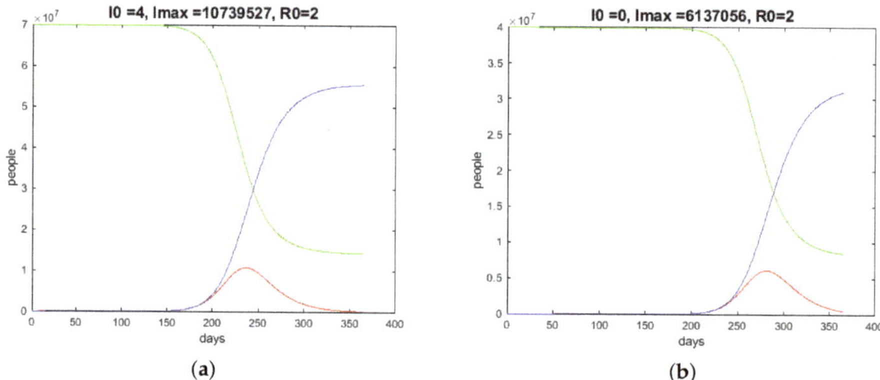

Figure 5. German and Spanish results over one year; S—green, I—red, R—blue. (**a**) German progression over one year with reduced β, starting at the end of January 2020; (**b**) Spanish progression over one year with reduced β, starting at the end of January 2020.

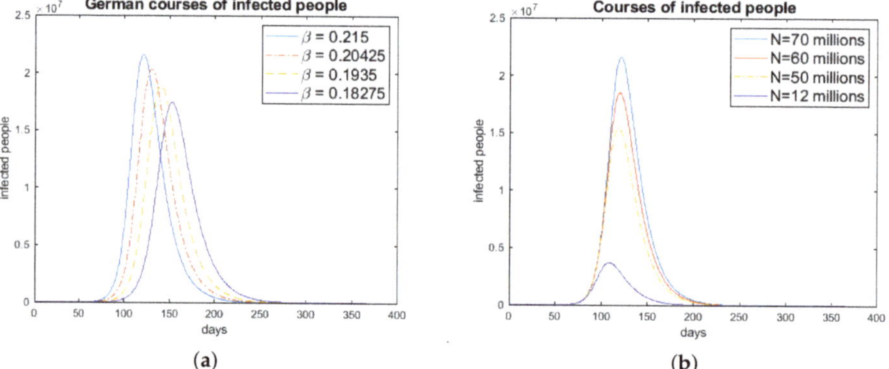

Figure 6. One year runs depending on β and N. (**a**) German succession of one year depending on a β-variatio; (**b**) Result of one year depending on N ($\beta = 0.215$).

5. Looking for Other Strategies of a Temporary Lockdowns and Extensive Social Distancing

In all countries concerned by the COVID-19 pandemic, a lockdown of social life has been discussed. In Germany, the lockdown started on 16 March 2020. The effects of social distancing to decrease the infection rate can be modeled by a modification of the SIR model. Now, I consider κ in the equation system (1)–(3) as a time-dependent function (instead of $\kappa = 1$ in the original SIR model).

κ is a function with values in $[0,1]$. For example,

$$\kappa(t) = \begin{cases} 0.5 & \text{for } t_0 \leq t \leq t_1 \\ 1 & \text{for } t > t_1, \; t < t_0 \end{cases}$$

indicates a reduction of the infection rate of 50% in the period $[t_0, t_1]$ ($\Delta_t = t_1 - t_0$ is the duration of the temporary lockdown in days). A good choice of t_0 and t_k will be complicated.

If I respect the chosen starting day of the German lockdown, 16th of March 2020 (this conforms to the 46th day of the relational year started at the end of January 2020), and I work[5] with

$$\kappa(t) = \begin{cases} 0.2 & \text{for } 46 \leq t \leq 76 \\ 1 & \text{for } t > 76, \ t < 46, \end{cases}$$

then I get the result pictured in Figure 7a.

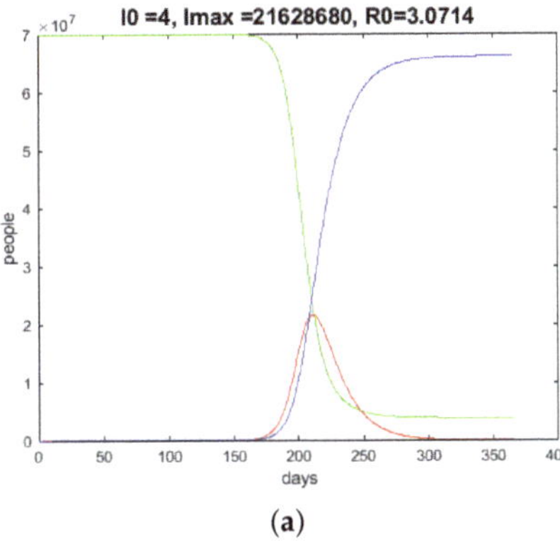

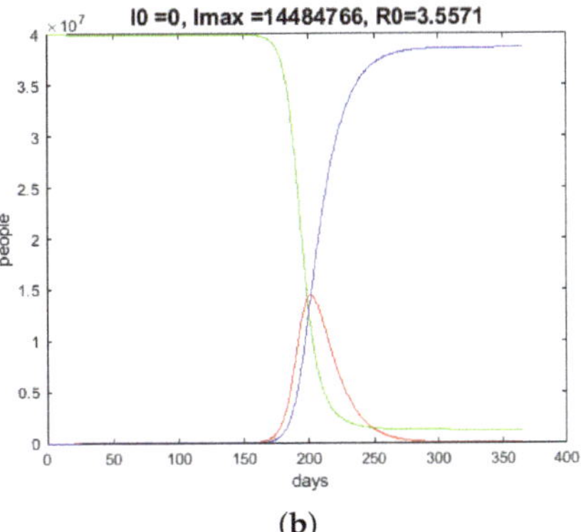

(a) (b)

Figure 7. Results with lockdowns; *S*—green, *I*—red, *R*—blue; 30 days lockdown, starting on 16 March 2020. (**a**) German progression over one year, starting at the end of January 2020; (**b**) Spanish progression over one year, starting at the end of January 2020.

The numerical tests showed that a very early start of the lockdown, resulting in a reduction of the infection rate β, causes the typical Gaussian curve to be delayed by I; however, the amplitude (maximum value of I) does not really change.

It is known from other pandemics, such as the Spanish flu ([11,12]) or the swine flu, that the development of the number of infected people looks like a Gaussian curve. The interesting points in time are those where the acceleration of the numbers of infected people increases or decreases, respectively.

These are the points in time where the curve of I changes from a convex to a concave behavior or vice versa. The convexity or concavity can be controlled by the second derivative of $I(t)$.

Let us consider Equation (2). By differentiation of (2) and the use of (1), I get

$$\begin{aligned} \frac{d^2 I}{dt^2} &= \frac{\beta}{N}\frac{dS}{dt}I + \frac{\beta}{N}S\frac{dI}{dt} - \gamma\frac{dI}{dt} \\ &= -\frac{\beta^2}{N}SI^2 + (\frac{\beta S}{N} - \gamma)(\frac{\beta S}{N} - \gamma)I \\ &= [(\frac{\beta S}{N} - \gamma)^2 - (\frac{\beta}{N})^2 SI]I . \end{aligned}$$

With that, the *I*-curve will change from convex to concave if the relation

$$(\frac{\beta S}{N} - \gamma)^2 - (\frac{\beta}{N})^2 SI < 0 \Longleftrightarrow I > \frac{(\frac{\beta S}{N} - \gamma)^2 N^2}{\beta^2 S} \tag{7}$$

---

[5] I will understand 20% of normality by a lockdown, this means $\kappa = 0.2$.

is valid. The switching time follows

$$t_0 = \min_t \{ t > 0, \; I(t) > (\frac{\beta S(t)}{N} - \gamma)^2 N^2) / (\beta^2 S(t)) \} \,. \tag{8}$$

A lockdown starting at t_0 (assigning $\beta^* = \kappa\beta$, $\kappa \in [0, 1[$) up to a point in time $t_1 = t_0 + \Delta_t$, with Δ_t as the duration of the lockdown in days, will be denoted as a dynamical lockdown (for $t > t_1$, β^* is reset to the original value β).

t_0 indicates the point in time up to which the growth rate increases and after which it decreases. Figure 8a shows the result of such a computation of a dynamical 30-days lockdown. I obtained $t_0 = 108$ ($\kappa = 0.2$). The result is significant. In Figure 9a, a typical behavior of $\frac{d^2 I}{dt^2}$ is plotted (in Figure 9b, $\frac{d^2 I}{dt^2}$ in the dynamical lockdown case).

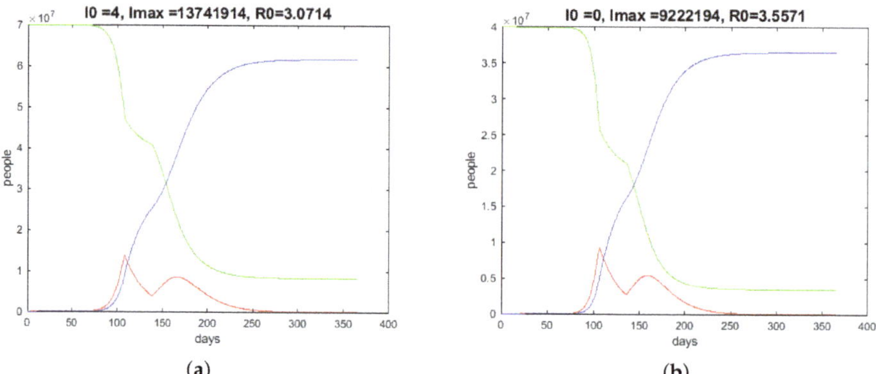

(**a**) (**b**)

Figure 8. Results over one year; *S*—green, *I*—red, *R*—blue. (**a**) German progression over one year, starting at the end of January 2020, dynamical lockdown; (**b**) Spanish progression over one year, starting at the end of March 2020, dynamical lockdown.

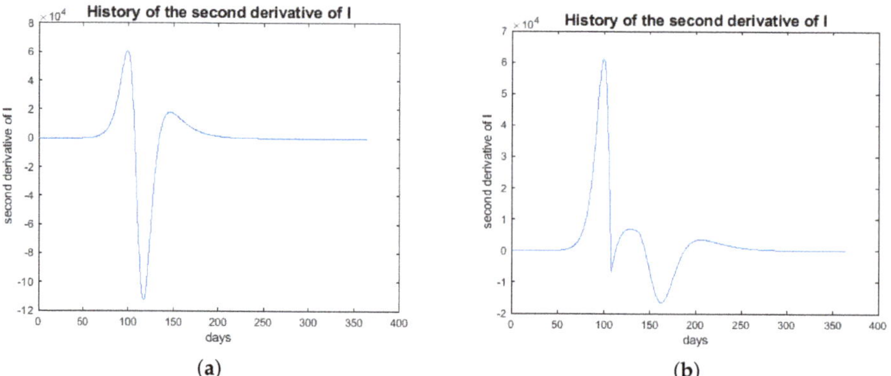

(**a**) (**b**)

Figure 9. Typical history of the second derivatives of *I*. (**a**) History of the second derivative of *I*; (**b**) History of the second derivative of *I* with dynamical lockdown.

The result of a dynamical 30-day lockdown for Spain is shown in Figure 8b, where I found $t_0 = 106$ ($\kappa = 0.2$).

Data from China and South Korea suggest that the group of infected people with an age of 70 or more is of a magnitude of 10%. This group has a significantly higher mortality rate than the rest of the infected people. Thus, I can presume that, as a high-risk group, $\alpha = 10\%$ of I must be especially sheltered and possibly medicated very intensively.

Figure 10a shows the German time history of the above-defined high-risk group with a dynamical lockdown with $\kappa = 0.2$ compared to the regime without social distancing. The maximum number of infected people decreases from approximately 1.7 million people to 0.8 million in the case of the lockdown (30-day lockdown).

This result proves the usefulness of a lockdown or strict social distancing during an epidemic disease. I observe a flattening of the infection curve as requested by politicians and health professionals. With strict social distancing for a limited time, one can save time to find vaccines and time to improve the possibilities of helping high-risk people in hospitals.

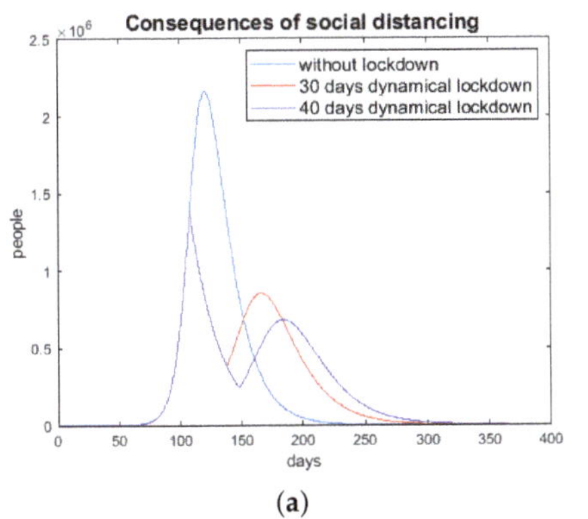

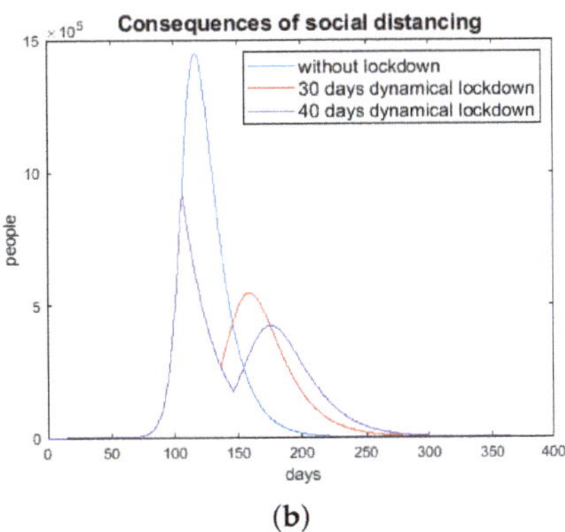

Figure 10. History of the high-risk groups depending on a dynamical lockdown. (**a**) German history; (**b**) Spanish history.

To see the influence of social distancing, I look at the Spanish situation without a lockdown and with a dynamical lockdown of 30 days in Figure 10b ($\kappa = 0.2$) for the 10% that includes high-risk people.

The computations with the SIR model show that the limited social distancing with a lockdown will be successful with a start after a time greater or equal to t_0, found by the evaluation of the second derivative of I (formula (8)). If the limited lockdown is started at a time less then t_0, the effect of such social distancing is not significant.

Bavaria is one of the origins of the German pandemic and is still under strict observation. Therefore, I will consider the simulation results for this German hotspot. I use $\beta = 0.215$ and $N = 12$ million as parameters. In Figure 11 the results for one year without and with lockdowns are shown.

In Figure 12a, the consequences of a 40-day social distancing/dynamical lockdown for the development of the number of high-risk infected people are shown. Because of the increasing number of infected people after the 40-day lockdown, I simulated a step-wise return to normality. After the 40-day lockdown, two 40-day periods follow with 60% and 80% of normality, respectively. The result of this simulation is shown in Figure 12b.

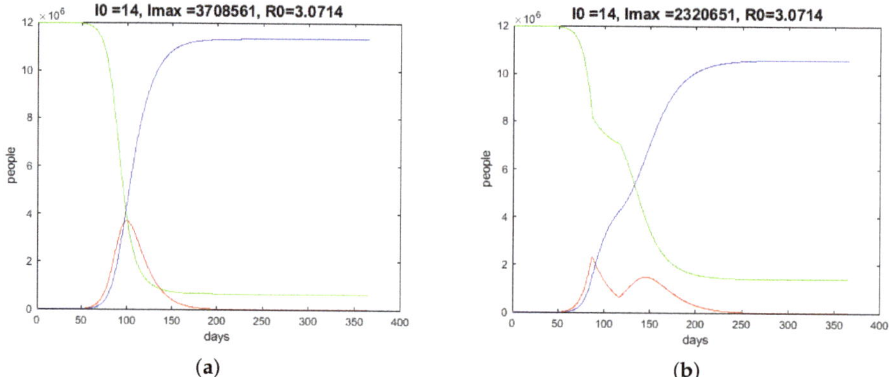

Figure 11. Bavarian one-year results; *S*—green, *I*—red, *R*—blue. (**a**) Bavarian one-year progression without lockdown; (**b**) Bavarian one-year progression with lockdown.

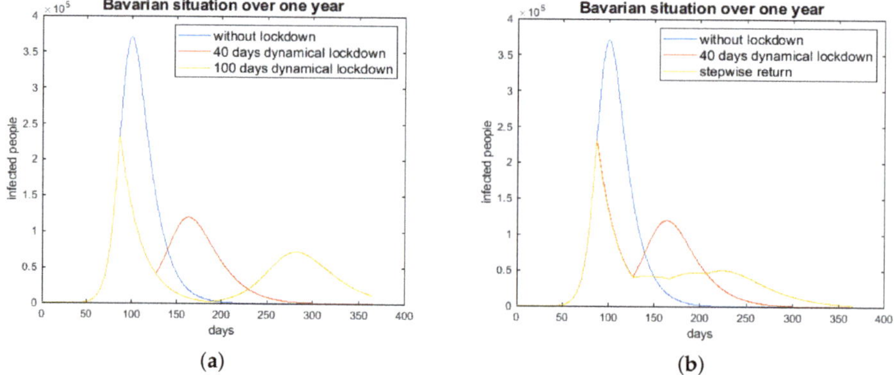

Figure 12. Bavarian one-year results for the high-risk people. (**a**) Bavarian one-year progression. (**b**) Bavarian one-year progression, with the green curve representing the step-wise return to normality.

The results for Bavaria with the considered step-wise lockdown can be passed to other regions or countries with pandemics. Such a strategy should be preferred instead of a complete return to normality after rigorous social distancing.

If I write (2) of the SIR model in the form

$$\frac{dI}{dt} = \left(\kappa\beta\frac{S}{N} - \gamma\right)I,$$

I realize that the number of infected people decreases if

$$\kappa\beta\frac{S}{N} - \gamma < 0 \iff S < N\frac{\gamma}{\kappa\beta} \tag{9}$$

is complied with. The relation (9) shows that there are possibilities for the reduction of infected people to be inverted and the medical burden to be reduced.

(a)   The first possibility to decrease the number of infected people is the reduction of the infection rate $\kappa\beta$. This can be achieved through strict lockdowns, social distancing at appropriate times, or rigid sanitary measures.

(b)   The second way consists of the reduction of the stock of the species S. This can be achieved through immunization or vaccination.

(c)   The isolation of high-risk people (70 years and older) is another possibility for the reduction of the number of infected people. In addition, positive tests for antibodies reduce the stock of susceptible persons.

If there is quantitative information on the isolation of infected people through quarantine, the SIR model can be extended by a species X, which quantifies symptomatic and quarantined infected people. This was considered in [2] for the Chinese province of Hubei.

6. Discussion and Conclusions

In this paper, I used a modified SIR model to describe the progression of the COVID-19 pandemic. I find that the timing of the lockdown is crucial in the progression of a pandemic. It could be shown that a very early start of limited social distancing measures of a period of Δt days leads only to a displacement of the climax of the pandemic, but not really to an efficient flattening of the curve of the number of infected people.

The intervention measures are more efficient, and one can observe a descent in the number of infected people if the social distancing is started beyond the dynamical lockdown time t_0. However, in this case, a second bump of the curve of infected people will also occur. A stepwise return to normality turned out to be the most efficient way to overcome the climax of a pandemic.

For the calibration of the SIR model, i.e., the evaluation of the parameter β, the non-linear regression comes up with significantly better results than the log–linear regression. This is evident with the comparison of the graphs of the evaluated exponential functions.

It must be noted again that the parameters β and κ were guessed very roughly. In addition, the percentage representing the group of high-risk people, α, is possibly overestimated. Depending on the capabilities and performance of the health systems of the respective countries, those parameters may look different. The interpretation of κ as a random variable is thinkable, too.

I have to point to the second bump in the progression of the number of infected people as an important issue of limited lockdowns. This must be respected in all decisions of physicians and politicians in connection with the handling of the pandemic. The simulations for Bavaria pictured in Figure 12 show that there are return strategies that can reduce further ramps of the progression of the number of infected people.

In conclusion, it must be said that the results of the simulations using the SIR model describe, in a way, the worst case. A lot of interventions made by politicians and physicians can disturb the progression of the pandemic in a positive way. However, not all measures and interventions can be described by SIR-type models. This allows the conjecture that the real pandemic will be weaker than the simulation results of the model.

Funding: This research received no external funding.

Conflicts of Interest: The author declares no conflicts of interest.

References

1.   Kantner, M.; Koprucki, T. Beyond just "flattening the curve": Optimal control of epidemics with purely non-pharmaceutical interventions. *arXiv* **2020**, arXiv:2004.0.471.

2.   Maier, B.; Brockmann, D. Effective containment explains subexponential growth in recent confirmed COVID-19 cases in China. *Science* **2020**, *368*, 742–746. [CrossRef] [PubMed]

3.   Saad-Roy, C.M.; Wingreen, N.S.; Levin, S.A.; Grenfell, B.T. Dynamics in a simple evolutionary-epidemiological model for the evolution of an initial asymptomatic infection stage. *Proc. Natl. Acad. Sci. USA* **2020**, *117*, 11541–11550. [CrossRef] [PubMed]

4.   Ndairou, F.; Area, I.; Nieto, J.J.; Torres, D.F. Mathematical modeling of COVID-19 transmission dynamics with a case study of Wuhan. *Chaos Solitons Fractals* **2020**, *135*. [CrossRef] [PubMed]

5. Barbarossa, M.V.; Fuhrmann, J.; Meinke, J.H.; Krieg, S.; Varma, H.V.; Castelletti, N.; Lippert, T. A first study on the impact of current and future control measures on the spread of COVID-19 in Germany. *medRxiv* **2020**. [CrossRef]

6. Kermack, W.; McKendrick, A. Contributions to the mathematical theory of epidemics—I. *Bull. Math. Biol.* **1991**, *53*, 1–2.

7. Bulletins of the European Centre for Disease Prevention and Control. 2020. Available online: https://www.ecdc.europa.eu/en/geographical-distribution-2019-ncov-cases (accessed on 21 March 2020).

8. Bulletin of Istituto Superiore della Sanita, Che cos'e R0 e Perche e Importante. 2020. Available online: https://www.epicentro.iss.it/en/coronavirus/sars-cov-2-integrated-surveillance-data (accessed on 21 March 2020).

9. Bulletins of the John Hopkins University of World-Wide Corona Data. 2020. Available online: https://www.jhu.edu (accessed on 21 March 2020).

10. Streeck, H.; Schulte, B.; Kuemmerer, B.; Richter, E.; Höller, T.; Fuhrmann, C.; Bartok, E.; Dolscheid, R.; Berger, M.; Wessendorf, L.; et al. Infection fatality rate of SARS-CoV-2 infection in a German community with a super-spreading event. *medRxiv* **2020**. [CrossRef]

11. Barry, J. The site of origin of the 1918 influenza pandemic and its public health implications. *J. Transl. Med.* **2004**, *2*, 1–4. [CrossRef] [PubMed]

12. Moradian, N.; Ochs, H.D.; Sedikies, C.; Hamblin, M.R.; Camargo, C.A.; Martinez, J.A.; Biamonte, J.D.; Abdollahi, M.; Torres, P.J.; Nieto, J.J.; et al. The urgent need for integrated science to fight COVID-19 pandemic and beyond. *J. Transl. Med.* **2020**, *18*, 205. [CrossRef] [PubMed]

Article

COVID-19 Case Rates in the UK: Modelling Uncertainties as Lockdown Lifts

Claire Brereton [1],* and Matteo Pedercini [2]

[1]   Child Health Research Centre, University of Queensland, Brisbane, QLD 4101, Australia
[2]   Millennium Institute, 2200 Pennsylvania Ave NW, Washington, DC 20037, USA; mp@millennium-institute.org
*   Correspondence: claire.brereton@uq.edu.au; Tel.: +61-419-901-107

Abstract: Background: The UK was one of the countries worst affected by the COVID-19 pandemic in Europe. A strict lockdown from early 2021 combined with an aggressive vaccination programme enabled a gradual easing of lockdown measures to be introduced whilst both deaths and reported case numbers reduced to less than 3% of their peak. The emergence of the Delta variant in April 2021 has reversed this trend, and the UK is once again experiencing surging cases, albeit with reduced average severity due to the success of the vaccination rollout. This study presents the results of a modelling exercise which simulates the progression of the pandemic in the UK through projection of daily case numbers as lockdown lifts. Methods: A simulation model based on the Susceptible-Exposed-Infected-Recovered structure was built. A timeline of UK lockdown measures was used to simulate the changing restrictions. The model was tailored for the UK, with some values set based on research and others obtained through calibration against 16 months of historical data. Results: The model projects that if lockdown restrictions are lifted in July 2021, UK COVID-19 cases will peak at hundreds of thousands daily in most viable scenarios, reducing in late 2021 as immunity acquired through both vaccination and infection reduces the susceptible population percentage. Further lockdown measures can be used to reduce daily cases. Other than the ever-present threat of the emergence of new variants, the most significant unknown factors affecting the profile of the pandemic in the UK are the length and strength of immunity, with daily peak cases over 50% higher if immunity lasts 8 months compared to 12 months. Another significant factor is the percentage of unreported cases. The reduced case severity associated with vaccination may lead to a higher proportion of unreported mild or asymptomatic cases, meaning that unmanaged infections resulting from unknown cases will continue to be a major source of infection. Conclusions: Further research into the length and strength of both recovered and vaccinated COVID-19 immunity is critical to delivering more accurate projections from models, thus enabling more finely tuned policy decisions. The model presented in this article, whilst by no means perfect, aims to contribute to greater transparency of the modelling process, which can only increase trust between policy makers, journalists and the general public.

Keywords: COVID-19; UK; vaccination; immunity; policy; system dynamics; modelling; uncertainty

Citation: Brereton, C.; Pedercini, M. COVID-19 Case Rates in the UK: Modelling Uncertainties as Lockdown Lifts. *Systems* **2021**, *9*, 60. https://doi.org/10.3390/systems9030060

Academic Editors: Oz Sahin and Russell Richards

Received: 10 May 2021
Accepted: 21 July 2021
Published: 6 August 2021

Publisher's Note: MDPI stays neutral with regard to jurisdictional claims in published maps and institutional affiliations.

1. Introduction

The COVID-19 pandemic is an unprecedented global crisis. The unusual nature of the SARS-CoV-2 virus, which can be deadly for one person whilst having no symptoms for another, was misunderstood by scientists and policy makers during the early stages of the pandemic, leading to underestimation of case numbers and focus on control of symptomatic infections [1]. Modelling studies [2,3] and research on the prevalence of COVID-19 antibodies in the UK population [4] indicated early on that confirmed cases were less than half of true infection estimates, and this reality is reflected in global pandemic planning guidance [5] and in the continuing use of measures such as lockdowns, which restrict social contact irrespective of known infection status across an entire population.

The United Kingdom (UK) was one of the countries worst affected by COVID-19 in the developed world, characterized by a slow initial response, lack of border controls, changing regional guidance and ease of movement between regions [6]. The UK is made up of four countries—England, Scotland, Wales and Northern Ireland, each with the autonomy to establish their own COVID-19 controls—but as 84% of the population resides in England, the profile of the pandemic in England and the measures taken there are the most significant driver of the UK's COVID-19 statistics. The escalating number of cases and deaths in the UK led to their being the first country to give authorisation for emergency use of the Pfizer/BioNTech (PB) vaccine. The vaccination programme started on 8 December 2020 and committed funds for an initial 30 million doses [7]. The AstraZeneca (AZ) vaccine was authorised on the same basis for rollout commencing 4 January 2021, with 100 million doses ordered. These vaccines delivered the capability to immunise 50 million people, effectively covering the entire eligible population of the UK for two doses each [8]. By end June 2021, 78 million vaccinations had been administered, with 33 million people fully vaccinated. The Moderna vaccine was also approved by the UK Government [9], and in mid-April 2021, it started rolling out to under 30 year olds as an alternative to AZ.

Mass vaccination has two main objectives: to protect individuals from death and severe illness and to increase the number of immune individuals to the point where enough people are protected from the virus to protect the population as a whole (herd immunity). For both vaccinated and recovered individuals, the longevity of protection from infection and the degree of protection conferred are still uncertain. The level of population protection required for herd immunity in the UK, or any other country, has been estimated but is as yet unknown.

As the COVID-19 pandemic has evolved, new strains have emerged, and in the UK, the Alpha variant and Delta variant have successively become dominant. Each of these strains have been more infectious than their predecessors, increasing the challenges to health systems.

Modelling studies have reached a new level of public health importance in 2020/2021 as policy makers have seen their value for predicting and analysing the future progression of the COVID-19 pandemic and allowing a comparison of interventions and policy decisions. There are broadly two modelling approaches being used. Mechanistic (dynamic) models such as the Imperial College London (ICL) model [10] reflect the underlying transmission process and contain non-linear feedback loops and delays, enabling longer term projection and inference of the results of changing assumptions or scenarios [11]. Statistical models, for example the Institute for Health Metrics and Evaluation model [12], use regression based or machine learning methods. These models do not account for how transmission occurs and are therefore not so well suited for long term projections about epidemiological dynamics. The Scientific Advisory Group for Emergencies (SAGE) in the UK uses a number of models to inform its advice [13]. In order to support a broad public debate on the upcoming precautionary measures against COVID-19, we develop a simulation model with three purposes:

1. to investigate the likely effects of lockdown easing on the UK pandemic, exploring the remaining uncertainties on vaccine efficacy and post-infection immunity;
2. to estimate the unknown proportion of COVID-19 cases in the UK and the role of unknown cases in the spread of the disease;
3. to increase the transparency of the modelling and analysis process, by focusing on containing the model detail complexity and clearly establishing the implications of different assumptions.

2. Background

2.1. Recovered and Post-Vaccination Immunity

As the COVID-19 epidemic continues in the UK, recovered population immunity is building. There is growing consensus amongst researchers that recovered immunity will not be lifelong and may be ineffective against new strains. Seasonal coronaviruses such as COVID-19, which infect mucosal surfaces and do not have a viremic phase, typically result in antibody responses that are detected for months or a few years [14]. Estimates of the longevity of recovered immunity range from at least 5 months to more than 12 months [15–17]. The longevity and level of protection of post-vaccination immunity is not necessarily the same as that of recovered immunity and will also become better understood with elapsed time, as will the protection which it gives against emerging variants. The first studies specific to COVID-19 reported that in the short term, recovery from infection gave 83% protection (95% CI 76–87%) from reinfection for at least 5 months [18,19]. Results from newer UK population research released in April 2021 showed 70% (95% CI 62–77%) protection from reinfection after either infection or vaccination [20]. Clinical trials continue to investigate vaccine efficacy, the protective effect of past infection and the effectiveness of both vaccines and past infection against emerging COVID-19 strains.

2.2. Transmissibility after Vaccination

Vaccine efficacy has three components: prevention of infection, reduction of disease severity and prevention of transmission [21]. Results from clinical trials focus on prevention and severity of infection, which is directly measurable, rather than on prevention of transmission. For this study, the relevant component of vaccine efficacy is its effectiveness in protecting against onwards transmission of the virus. Research shows that the UK's vaccination programme has resulted not only in protection from infection but also in a lower viral burden if infected, leading to a much higher proportion of asymptomatic and mild infections. Comparison of viral burden in vaccinated and unvaccinated groups shows a 65% decrease three weeks after one dose of either AZ or PB, and a 70% decrease 1 week after a second dose [20]. Viral burden can be used as a proxy for post-vaccination transmissibility decrease, which is not directly measurable.

2.3. Known, Unknown and Asymptomatic Cases

Asymptomatic transmission is recognised as a significant contributor to the COVID-19 pandemic, both from pre-symptomatic individuals and from those who never develop symptoms [22,23]. The effect of vaccines in reducing the severity to asymptomatic or mild disease may also mean that more cases go undetected in the community, contributing to increased transmission [24]. At least 50% of new infections are estimated to have originated from exposure to individuals with infection but without symptoms [25]. Evidence suggests a 42% lower transmission rate for asymptomatic cases [26,27]. It is broadly acknowledged that there is massive global under-reporting of symptomatic COVID-19 cases for many reasons ranging from perception of low personal risk from COVID-19 infection to lack of trust in health services, lack of testing capacity and a desire to avoid the negative consequences of enforced isolation [28]. The unknown proportion of cases is thus likely to be higher than the truly asymptomatic proportion and the modelling exercise uses optimisation techniques to estimate this unknown proportion.

3. Method and Data Sources

3.1. Model Development

We developed a dynamic model of the COVID-19 pandemic based on the established Susceptible-Exposed-Infected-Recovered (SEIR) compartmental infectious disease model structure [29]. The model, shown in Figure 1, was constructed using Stella Architect software supplied by isee systems, Lebanon, NH, USA.

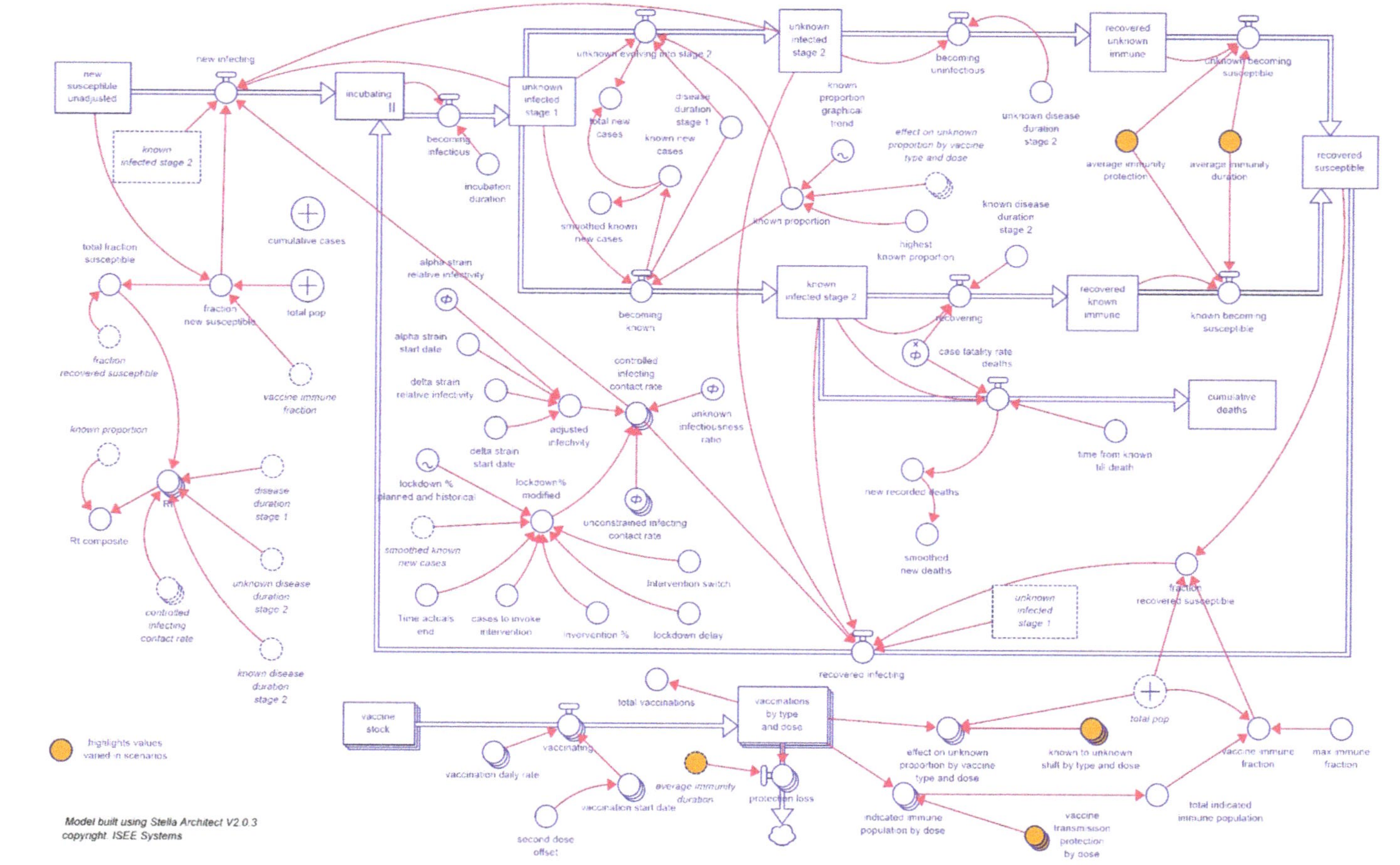

Figure 1. SEIR model of COVID-19 pandemic in UK.

The SEIR system structure is based on a reinforcing feedback loop of exponentially growing infections over time, balanced by an eventual reduction of susceptible individuals due to death or increasing population immunity. Speed of transmission is tracked by the calculated reproduction number, R_t, with daily case numbers reducing when R_t falls below 1 (R_0, initial reproduction number, is often used incorrectly in place of R_t).

The model includes the effects of the social distancing and infection spread measures used to control the spread of COVID-19. Infections are classified as known or unknown, with the parameters associated with contact rates given different values depending on known/unknown status. The effects of a vaccination programme, which reduces the susceptible population, and the effects of recovered immunity drop-off [30], which increases the susceptible population, are also included.

The model consists of stocks, flows and auxiliary variables including intermediate calculations for the determination of flows. Stocks represent levels or state variables, including the numbers of people in the different infectious states or the numbers of vaccine doses available; these are represented by rectangles. Flows represent the rates at which people and doses transition between states and are represented by valve symbols. These rates are determined by time constants or probability estimates of moving to one state or another. The model captures the fundamental drivers of the COVID-19 pandemic and does not provide spatial or individual-level disaggregation. Its lack of detail complexity is meant to provide transparency in the modelling and analysis process, whilst allowing the exploration of a broad range of alternative scenarios.

The model runs from 1 February 2020, when the total population is susceptible, to 31 December 2021, with a time step of 6 h. Individuals acquire the infection, incubate the disease during an initial latent period and then become infectious. Each stage introduces a delay into the system. An individual's infectious state is at first unknown, then, as the disease becomes symptomatic, it becomes known in a proportion of the infected population. Some individuals' infectious state is never known to health authorities, either because they are asymptomatic or because they do not recognize or wish to disclose their symptoms for various reasons. Most infected individuals recover, with a proportion of known infected individuals dying. Recovered individuals acquire a level of protective immunity, which reduces the susceptible population. The model also projects the effects of potential future UK Government interventions by simulating increased lockdowns when known daily cases rise above threshold levels. All equations, auxiliary variable values and initial values of stocks are listed in Supplementary Materials Table S1.

3.2. Model Data Sources

The infection rate in the model is calculated from the susceptible population and the daily infecting contact rate, which is affected by social distancing, hygiene and lockdown measures and is significantly lower for known infected individuals. Infectivity in the model increases from 5 December 2020 and again from 13 April 2021, reflecting the emergence of the 'UK variant' B.1.1.7, now known as the Alpha variant, which was measured as 35% more contagious (95% CI 2–69%) [31,32] and then the 'Delta variant', assumed to be twice as contagious as the original virus. The model uses data for the PB and AZ vaccines only, as the Moderna vaccine has not yet been deployed in quantity in the UK.

The values of the parameters used in the model, shown in Table 1, were established in two ways:

1. For parameters where reliable data was available from published research, e.g., virus incubation time, the median values from the research were used;
2. For parameters where data was either unavailable or considered unreliable, the Powell optimisation method was used to calibrate the model and confirm a narrow spread of 95% confidence intervals.

Table 1. Major parameter values used in model.

Parameter	Value	Unit	Source
Incubation duration (non-infectious latent period)	3.5	Days	[33]
Disease duration stage 1 unknown	2	Days	[33,34]
Disease duration stage 2 known	8	Days	[33,34]
Disease duration stage 2 unknown	5	Days	[33,34]
Time from known disease till death	11	Days	[34]
Vaccine rollout speed PB/AZ	130,000/380,000	Doses/day	[35,36]
Vaccine protection against onwards transmission 21 days after dose 1 PB/AZ	65% [$]	-	[20]
Vaccine protection against onwards transmission 7 days after dose 2 PB/AZ	70% [$]	-	[20]
Length of immunity after vaccination or recovery	8 [$]	Months	[15]
Maximum population immunity	70%	-	[37]
Average immunity protection post recovery	70% [$]	-	[20]
Unknown infectiousness ratio *	72% [$]	-	[5,26,27,38–40] and model optimisation
Unconstrained infecting daily contact rate unknown	0.56 [$]	-	model optimisation
Unconstrained infecting daily contact rate known	0.14 [$]	-	model optimisation
Known proportion estimate February 2021	21% [$]	-	[2] and model optimisation
Relative infectivity after alpha variant identified	1.32 [$]	-	[32] and model optimisation
Relative infectivity after delta variant identified	2.0 [$]	-	[41]

* Starting point was the best estimate used by Center for Disease Control and Prevention based on multiple assumptions and conflicting research papers. [$] Value used for base case of model.

3.3. Lockdown Effectiveness Timeline Estimation

As social distancing and lockdowns have proven to be one of the most effective ways of combating the spread of the virus [42], a composite measure of lockdown effectiveness based on the timeline of the various restrictions and their easing measures was a key part of the model. This measure is known as the 'lockdown percentage'. It varies throughout the life of the model and measures the timeline of social distancing, mask wearing and movement restriction measures and varies between 0% and 100%, where 0% represents society with no restrictions in place and 100% a hypothetical total restriction scenario with no contact and therefore no transmission of the virus.

From January 2021, the UK Government implemented a set of country lockdown plans which specified staged step downs separated by a minimum of five weeks, with 7 day's notice of each change [43] to enable the observation of the data before proceeding. The dates of the most significant measures taken and the future plans [43] are shown in Table 2. The lockdown percentage timeline was estimated from this table and compared with data from a UK social distancing measures adherence study [44].

Table 2. Dates of significant measures.

Event	Date
First two UK COVID-19 cases confirmed	1 February 2020
UK Government Coronavirus action plan	3 March 2020
First COVID-19 death	3 March 2020
Contact tracing abandoned	12 March 2020
UK-wide lockdown effected	26 March 2020
Prime Minister admitted to hospital with COVID-19 symptoms	4 April 2020
COVID-19 alert levels system announced	1 May 2020
Lockdown eased, workers return, outdoor exercise with another	13 May 2020
Lockdown eased, non-essential shops reopen	15 June 2020
Restaurants and pubs open	4 July 2020
Restaurant 'eat out to help out' campaign	3 August 2020
One of every three cases in 20–29-year-olds, fast growth in younger people	7 September 2020
England—'Rule of Six' announced to curb social gatherings	14 September 2020
England—three-tier alert framework implemented	14 October 2020
Northern Ireland—4-week 'circuit breaker' lockdown starts	16 October 2020
Wales—3-week 'firebreak' lockdown starts	23 October 2020
Scotland—5-tier alert system starts	2 November 2020
England—4-week national lockdown starts at new tier 4	5 November 2020
New COVID-19 strain (Alpha variant) B.1.1.7 detected in UK	13 November 2020
England—4-week lockdown ends	3 December 2020
PB immunisation rollout starts	8 December 2020
London and Scotland, new tier 4 lockdown	20 December 2020
Christmas one day lockdown relaxation	25 December 2020
AZ immunization rollout starts	4 January 2021
England, Scotland—tier 5 lockdown to 22 February	6 January 2021
England—lockdown extended to 8 March	27 January 2021
Schools return	8 March 2021
Non-essential retail, outdoor hospitality and attractions reopen	12 April 2021
New COVID-19 strain (Delta variant) B.1.617.2 detected in UK	15 April 2021
Indoor hospitality and sporting events with limited capacity reopen	17 May 2021
Planned England and Scotland 'Freedom day' 21 June deferred to 19 July	14 June 2021
FUTURE CHANGES:	
England—mandatory mask rules lifted, nightclubs reopen, full capacity events	19 July 2021
Scotland—level zero, up to 10 people meet indoors, nightclubs remain closed	19 July 2021

3.4. Model Calibration and Optimisation

The model was calibrated against historical UK COVID-19 case, death and vaccination data up to 12 July 2021 sourced from Johns Hopkins University [36]. Calibration was done using an optimisation process to find the model variables which produced the best fit to the historical data. The variables which were used for optimisation were: the known and unknown infecting contact rates, the infectiousness ratio of unknown to known cases and the known proportion of cases. This optimisation produced the model 'base case' which was used as the starting point for varying uncertainties. Optimisation was also performed for differing immunity length scenarios. The relative infectivity of the Alpha variant and the Delta variant were calibrated by later optimisations.

After calibration, the following validation checks were performed:

- The 'new susceptible' and 'recovered susceptible' stocks in the model were validated against UK COVID-19 antibody prevalence studies to ensure that the population fraction of people with antibodies, who can be presumed to have recovered from COVID-19, aligns with the modelled fraction [4];
- Modelled UK case fatality rates were compared with historical data to ensure broad alignment [36];
- The reproduction number R_t, calculated by the model over time, was compared with studies of the initial R_0 and the ongoing COVID-19 R_t values to check consistency [45];

- The unknown infectiousness ratio was compared with previous research to ensure that it was at least as high as the estimated asymptomatic infectiousness ratio [26,27].

The major assumptions made in the model in addition to the assumed parameter values were:

- The relative infectivity increases at two points in time due to the new Alpha and Delta variants;
- Vaccination proceeds at a steady daily rate in all scenarios and is offered to the total eligible population irrespective of whether an individual is known to have recovered from COVID-19;
- The maximum achievable population immunity fraction of 70% is capped by ineligible population sectors (pregnant women and most children under 18), vaccine hesitancy [37] and logistical difficulties;
- The second dose of a vaccine is given 12 weeks after the first dose;
- The protective effect of the first dose of the vaccine is established 21 days after administration, and increased protection is established 7 days after the second dose;
- The average time lag between symptom onset and the reporting of a positive case to the data source is 4 days.

3.5. Uncertainty Modelling

Having established the model 'base case' through calibration and validation, uncertain parameters in the model were then varied between the 95% CIs reported in clinical trials, enabling the exploration of the effect on future daily case rates. A summary of the areas of uncertainty investigated is shown in Table 3.

Table 3. Scenarios simulated in model.

Scenario	Immunity Length Post Vaccination and Post Recovery	Protection from Infection Given by Recovered Immunity	Vaccine Protection 3 Weeks after 1st Dose	Vaccine Transmission Protection 1 Week after 2nd Dose	Future Known Proportion of Cases	Lockdown Characteristics
Base Case	8 months [15]	70% [20]	PB/AZ 65% [20]	PB/AZ 70% [20]	50%	-
Recovered immunity protection variations	8 months	62%/70%/87% [15–17,19,20]	PB/AZ 65%	PB/AZ 70%	50%	-
Vaccine protection variations	8 months	70%	PB/AZ 60%/65%/70% [20]	PB/AZ 62%/70%/77% [20]	50%	-
Known proportion of cases variations	8 months	70%	PB/AZ 65%	PB/AZ 70%	50%/37.5%/25%/12.5%	-
Lockdown sensitivity variations	8/12 months	70%	PB/AZ 65%	PB/AZ 70%	50%	Delays from 3.5 to 21 days, Case thresholds from 5000 to 25,000, Lockdown increase from 25% to 50%

There is no published research data available for post-vaccination immunity length, so this was assumed to be the same as post-recovery immunity. The proportion of known COVID-19 cases may reduce due to lowered disease severity; the model was run using values of 0%, 25%, 50% and 75% reduction in the absence of published research.

4. Results

4.1. Model Fit to Actuals

Figure 2 shows the reported historical and modelled 7-day averages for the UK's new known daily COVID-19 cases from 1 February 2020 to 12 July 2021. The error statistics calculations (R^2: = 0.97, RMSPE = 3.6% and Theil's inequality coefficient = 0.07) confirm a good fit of the simulated results to historical actuals. The lockdown percentage is represented as a black line with its scale on the right axis. The left axis shows the scales for the actual and modelled new known daily cases and deaths, with cases climbing to 60,000 in January 2021. The x-axis markings show the beginning of each month.

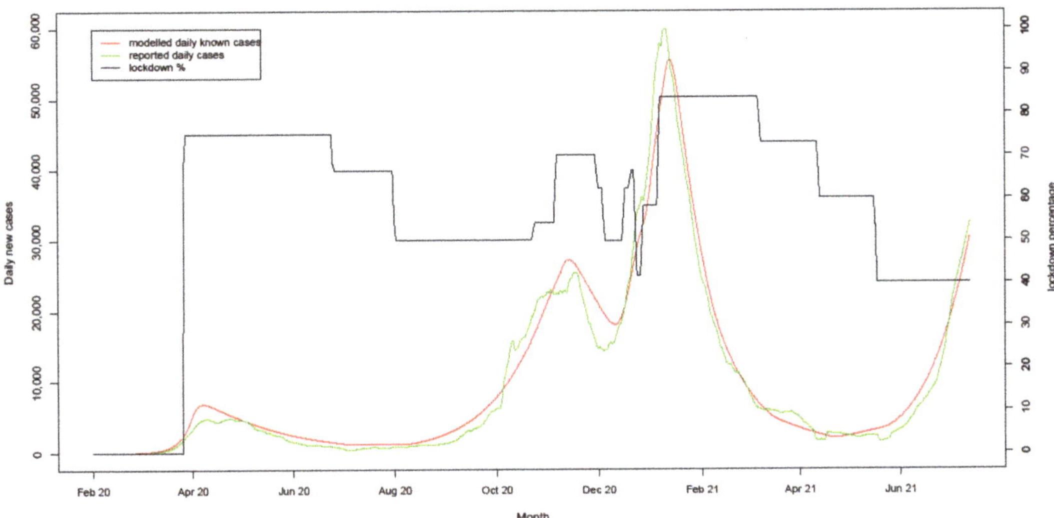

Figure 2. Daily UK reported COVID-19 cases 1 February 2020 to 12 July 2021.

The effect of the first UK-wide lockdown, which was estimated as 75% effective [44], can be seen in April 2020, with known case numbers peaking 16 days later. The gradual easing of the lockdown from 5 July 2020 resulted in an increase in known cases from August 2020, with the UK Government 'Eat out to help out' scheme estimated to have raised infection rates by 8 to 17% [46]. The lockdown percentage increased from mid-October 2020 in response to rising rates as the English tiered alert system started and Northern Ireland and Wales imposed 'firebreak lockdowns', followed by regional restrictions in Scotland and a four-week English lockdown starting 5 November in an attempt to reduce case numbers before the Christmas period. The effect of these consolidated lockdowns was to reduce the known case numbers from mid-November 2020 for 16 days, only for them to climb from 5 December 2020 onwards as the UK moved into its holiday period. The emergence of the more contagious Alpha variant in December 2020 accelerated the new case rate and made a strict lockdown in January 2021 necessary to contain the 'second wave'. The lockdown was effective in reducing cases, which peaked at 60,000 per day 12 days after the Christmas lockdown relaxation and then fell below 2000 per day in May 2021. However, the Delta variant, which became dominant in the UK in April 2021, combined with easing of lockdown restrictions in April and May, reversed the downwards trend and cases climbed to over 30,000 per day in July 2021.

The optimisation process described in Section 3.2 calculated a relative infectiousness value of 72% for unknown cases, which is in the range supported by the research [5]. The known proportion of 21% of cases at the end of January 2021 was also obtained through optimisation, assuming a logarithmic growth rate from the beginning of the model's timeframe. This is in the range supported by other models [2] and helps to explain why non-discriminatory lockdowns were adopted as the only effective means of controlling the spread of COVID-19 before vaccines were developed. The known proportion was assumed to increase to 50% by end March 2021 as cases fell, testing capability improved and self-testing became mandatory for certain professions, e.g., teaching. This assumption was validated by a comparison of reported cases against random population sampling.

4.2. Exploring Uncertainty

The scenarios identified in Table 3 were simulated by varying the selected variables whilst keeping other variables at 'base case' levels.

4.2.1. Uncertain Immunity Length

The 'base case' defined in Table 3 assumes 8 months average immunity, either after vaccination or recovery from infection [15], a 65% reduction in transmission protection after one dose, a 70% reduction after two doses of either the PB or AZ vaccine and 70% protection from reinfection after recovery from COVID-19 [20]. Research to date reports that immunity is likely to vary between 5 and 12 months [15–17,19], and Table 4 shows the simulated scenarios. Immunity against emerging variants may be different and is not accounted for in this model.

Table 4. Varying immunity scenarios.

Scenario	Immunity Length	Recovered Immunity Protection	Vaccine Protection 3 Weeks after 1st Dose	Vaccine Protection 1 Week after 2nd Dose	Future Known Cases
Immunity length variations	5/8/12 months	70%	PB/AZ 65%	PB/AZ 70%	50%

The model was run from 1 February 2020 to 31 December 2021 to simulate the 'base case' of 8 months immunity and shorter and longer average immunity lengths of 5 and 12 months. Figure 3 shows the projected daily known cases for the three scenarios, assuming a stepped lockdown percentage decrease from March 2021 onwards, which reduces to 20% in mid July 2021 according to the current UK Government timelines [43]. The figure of 20% assumes that some distancing restrictions are still in place until the end of 2021, that people will continue to exercise caution and that businesses will continue risk reduction policies such as disinfection and management of crowds.

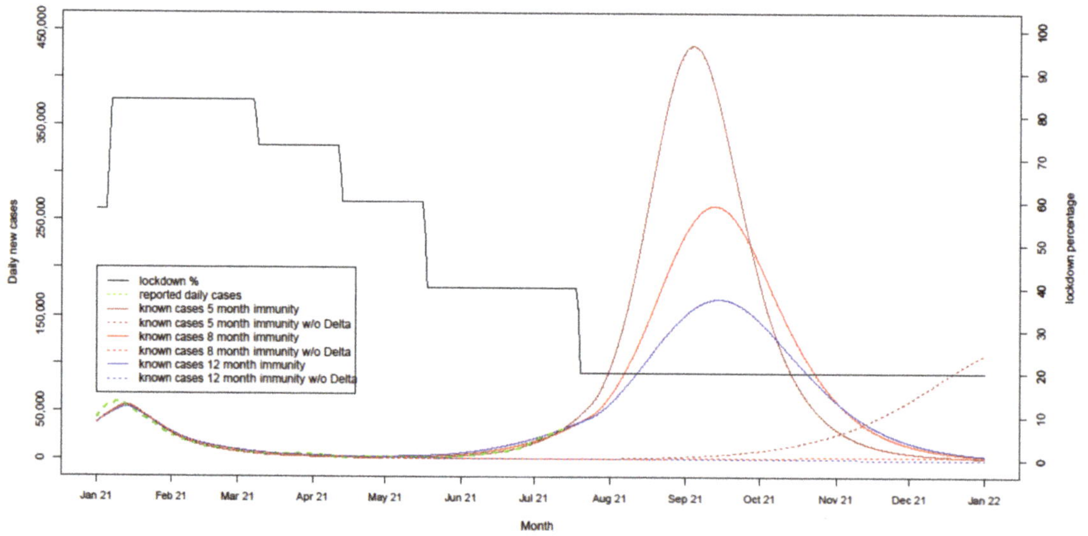

Figure 3. Daily UK COVID-19 cases projected to end 2021 with varying immunity lengths.

For the 'base case', the solid red line in Figure 3 shows the model's projection of a continuing rapid increase in known daily cases, driven by increased transmission opportunities and an increased susceptible population percentage as those infected in early 2021 lose their immunity. This peaks in September 2021 at 260,000 daily known cases when population immunity created by both vaccination and recovery from infection reduces the susceptible percentage and numbers start to fall. This projection is starkly different from the pre-Delta variant scenario, which is represented by the dotted red line. In this scenario, immunity from both vaccination and recovery would have contained daily known cases below 3000 from May 2021. Increasing the average immunity length to 12 months is projected to contain the surge to 160,000 daily known cases, peaking in October 2021. If immunity only lasts for 5 months, the surge is higher and a peak of 430,000 daily known cases is reached in August 2021. A 5-month immunity scenario assuming no Delta variant would also see cases rising more slowly, peaking in December 2021. The 5-month immunity scenarios, however, seem unlikely as actual known daily cases are not surging fast enough in July 2021 to align with the model's projections.

The results shown in Figure 3 are based on the assumption that from May 2021 onwards, 50% of cases continue to be detected due to increased testing capability. However, this detection rate may well be unachievable at these high case levels, in which case reported results would show lower numbers than those projected in the simulation.

4.2.2. Uncertain Immunity Effectiveness

Research has produced a range of effectiveness results and confidence intervals for both recovered and vaccinated immunity. Table 5 shows the varying immunity effectiveness scenarios simulated. The scenarios reflect the 95% CI range of post-vaccination and post-recovery immunity protection from the results of clinical research [20], assuming the 'base case' for other values [15–17,19,20]. The 95% CI ranges for recovered and vaccinated immunity are different, and this is reflected in the scenarios used. Figure 4 shows the modelled projections for these scenarios.

Table 5. Varying immunity effectiveness scenarios.

Scenario	Immunity Length	Recovered Immunity Protection	Vaccine Protection 3 Weeks after 1st Dose	Vaccine Protection 1 Week after 2nd Dose	Future Known Cases
Vaccine protection variations	8 months	70%	PB/AZ 60%/65%/70%	PB/AZ 62%/70%/77%	50%
Recovered immunity protection variations	8 months	62%/70%/87%	PB/AZ 65%	PB/AZ 70%	50%

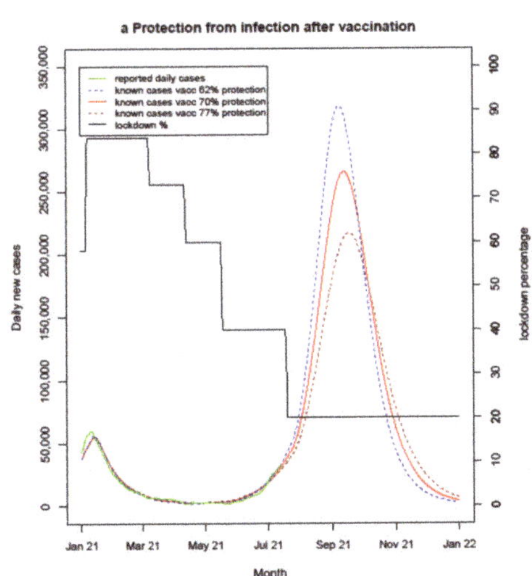

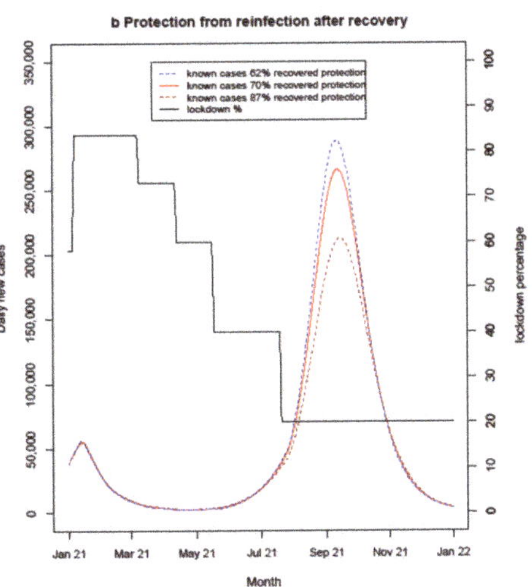

Figure 4. Daily known case projections with varying immunity protection.

Figure 4a projects that if post-vaccination protection from infection is at the lower boundary of 62% after two doses, known infections will build to 320,000 in September. Using the higher boundary of 77% protection after two doses, the model projects that known daily cases will peak at 210,000 before dropping as herd immunity from both vaccination and recovery reduces the susceptible percentage.

Figure 4b shows the projected range of known cases for recovered immunity variation. The model projects that the lower value of recovered immunity of 62% will result in a daily known case surge to 280,000 in September 2021, reducing to 215,000 with the higher value of 87%. As described in Section 4.2.1, 50% detection at these high daily case numbers may be unachievable, which would reduce the reported case peaks.

4.2.3. Uncertain Known Proportion

The results presented so far show only the known proportion of COVID-19 cases in the UK. As vaccination reduces not only the case numbers but also the average case severity, the unknown proportion may increase further as the proportion of mild or asymptomatic cases grows, even with increased ease and availability of testing. Table 6 shows the scenarios modelled.

Table 6. Varying known proportion scenarios.

Scenario	Immunity Length	Recovered Immunity Protection	Vaccine Protection 3 Weeks after 1st Dose	Vaccine Protection 1 Week after 2nd Dose	Future Known Cases
Known proportion variations	8 months	70%	PB/AZ 65%	PB/AZ 70%	50%/37.5%/25%/12.5%

Figure 5a,b project the daily known and total cases for 2021 for the 'base case' scenario with the percentage of known cases to unknown ranging from 50% to 12.5%. The base case assumes that 50% of cases are known.

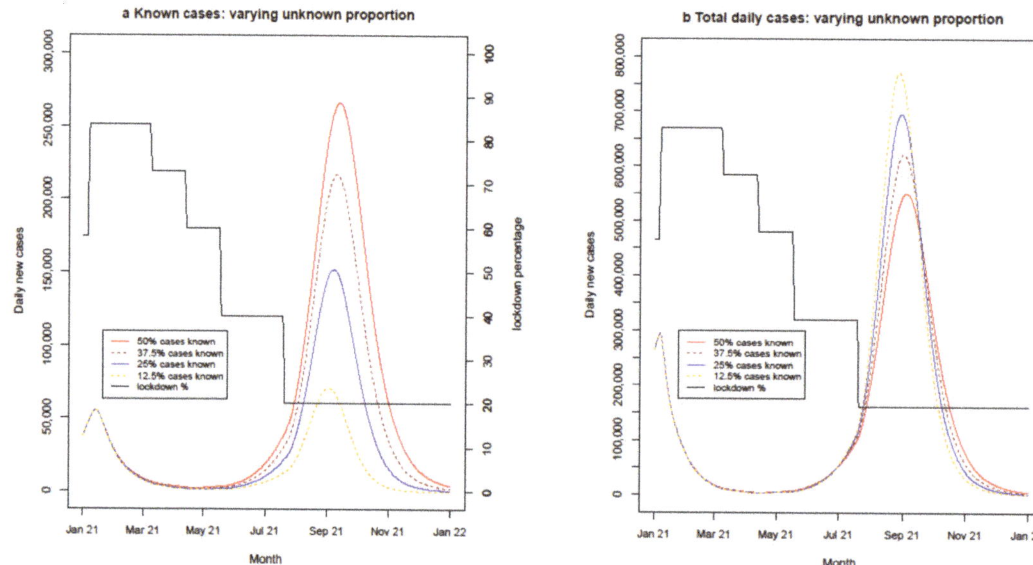

Figure 5. Daily known and unknown UK COVID-19 cases in 2021 with varying known proportion assumptions.

As expected, the projected known case numbers drop as the unknown proportion rises. The projected total cases would be expected to increase when a lower percentage of the cases are known because transmission is not being managed through isolation of infected individuals. However, because unknown cases are assumed to be less infectious and of a shorter duration than known cases [26,27], a 75% reduction in the proportion of known cases (from 50% to 12.5%) generates only a 40% increase in total case numbers.

4.2.4. Modelling the Effect of Interventions

The UK Government's planned landmark date of 21 June 2021, 'Freedom day', when masks could be removed and other significant restrictions would be lifted, was moved to 19 July as daily case numbers started to rise in May 2021 [47]. This rise, driven by the more transmissible Delta variant and the eased restrictions, raises the question of whether further lockdowns should be considered despite the increasing vaccination numbers. From the results shown in Figures 3–5, it can be seen that varying immunity length has a larger impact on case number projections than varying vaccination and recovered immunity protection within their likely ranges. Therefore, potential lockdown scenarios were explored with differing immunity length assumptions, as shown in Table 7.

Table 7. Varying lockdown initiation scenarios.

Scenario	Immunity Length	Recovered Immunity Protection	Vaccine Protection 1st Dose	Vaccine Protection 2nd Dose	Future Known Cases	Lockdown Daily Case Threshold	Lockdown%
Lockdown effects for varying immunity lengths	5/8/12 months	70%	PB/AZ 65%	PB/AZ 70%	7 days	50,000	20% addition

Figure 6a,b simulate the effects of a Government policy which reacts to daily known cases rising above 50,000 by increasing lockdown levels by 20%. The 20% is a theoretical number which could be made up of a number of different measures, e.g., self-isolation restrictions, masks, number limits. A 7-day reaction time is built into the simulation, in line with current Government policy.

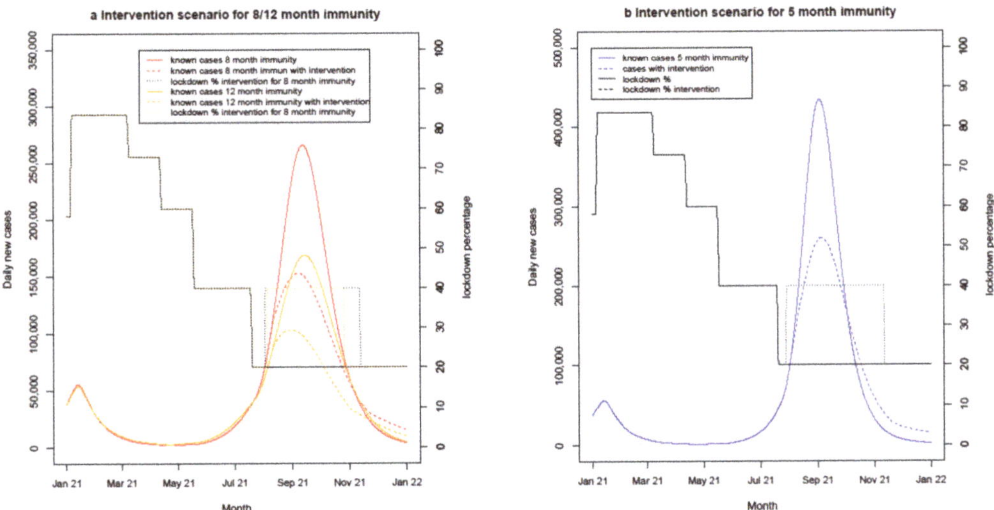

Figure 6. Lockdown interventions when cases rise above 50,000.

Figure 6a projects that for an 8-month immunity length, a 3-month-long return to the 40% lockdown level would be required from late July 2021 to return cases to below 50,000.

For a 12-month immunity length, a 2-month return to the 40% lockdown level would be required, starting at a similar time. Figure 6b projects that for a 5-month immunity length, the 50,000-case threshold will be breached in July and continuing lockdown at the July levels would reduce the peak daily numbers to 250,000 before they drop down in November 2021.

4.2.5. Lockdown Policy Sensitivities

The scenarios shown in Table 8 were used to simulate the sensitivity of the lockdown policy to the length of time before initiating lockdown.

Table 8. Varying lockdown initiation delay scenarios.

Scenario	Immunity Length	Recovered Immunity Protection	Vaccine Protection 1st Dose	Vaccine Protection 2nd Dose	Delay before Lockdown	Lockdown Daily Case Threshold	Lockdown%
Lockdown delay variations	8 months	70%	PB/AZ 65%	PB/AZ 70%	3.5, 7, 10.5, 14, 17.5, 21 days	5000	25% addition
	5 months	70%	PB/AZ 65%	PB/AZ 70%	3.5, 7, 10.5, 14, 17.5, 21 days	5000	25% addition

Figure 7 projects the results of varying the lockdown notice period between 3.5 and 21 days after known cases reach 50,000. Figure 7a shows that the 8-month immunity 'base case' with a 20% increase in lockdown percentage results in a shorter delay and a lower peak in cases. The highest peak is projected for the 21-day lead time. Figure 7b shows the same pattern for the 12-month immunity assumption, with maximum daily infections reaching 96,000 for a 3.5-day lead time and 136,000 for a 21-day lead time.

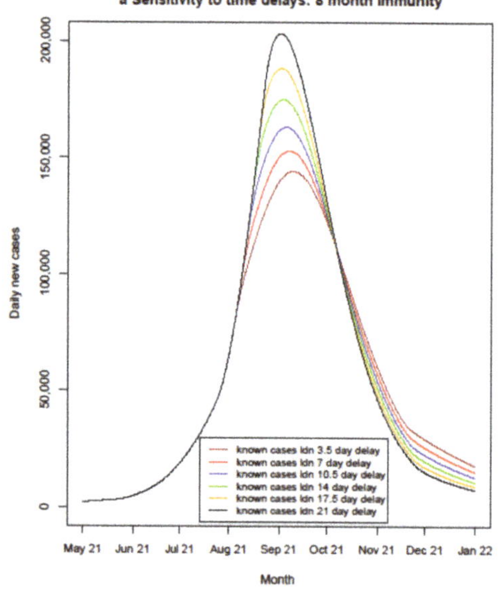

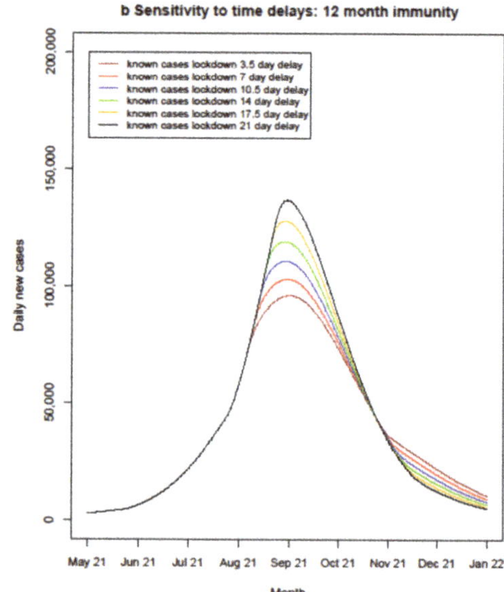

Figure 7. Effect of varying time to initiate lockdown.

The scenarios shown in Table 9 were used to simulate the sensitivity of the lockdown policy to the case threshold before initiating lockdown.

Table 9. Varying lockdown case threshold scenarios.

Scenario	Immunity Length	Recovered Immunity Protection	Vaccine Protection 1st Dose	Vaccine Protection 2nd Dose	Delay before Lockdown	Lockdown Daily Case Threshold	Lockdown%
Lockdown case threshold variations	8 months	70%	PB/AZ 65%	PB/AZ 70%	7 days	25,000, 50,000, 75,000, 100,000	20% addition
	12 months	70%	PB/AZ 65%	PB/AZ 70%	7 days	25,000, 50,000, 75,000, 100,000	20% addition

Figure 8 projects the results of varying the daily known case threshold for initiating lockdown between 25,000 and 100,000, assuming a 7-day lead time as per the current UK Government policy. Figure 8a shows that, for the 8-month immunity 'base case', the lower the case threshold, the lower the peak of daily cases. In all scenarios, cases fall rapidly as the susceptible percentage reduces due to increasing population immunity from the large numbers of recovered infections and vaccinations. Figure 8b shows the same pattern for the 12-month immunity scenario with lower peaks because of the greater level of retained recovered population immunity.

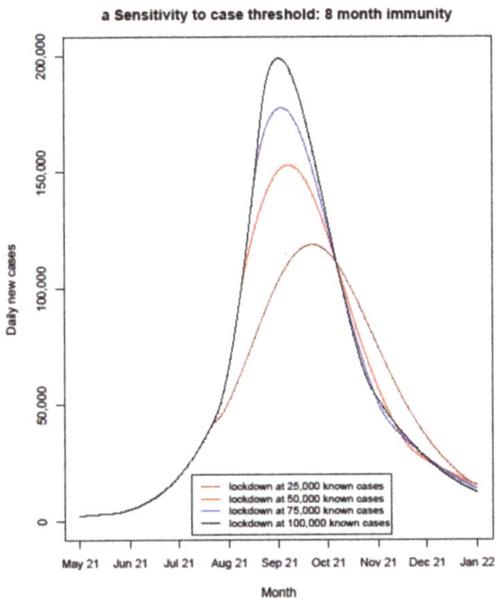

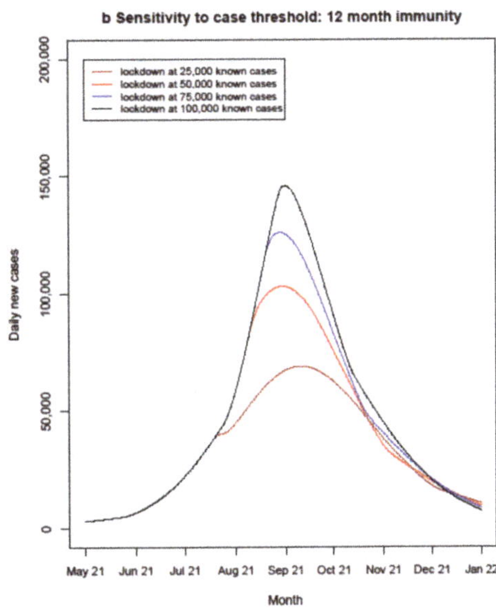

Figure 8. Effect of varying number of known cases required to initiate lockdown.

The model was used to simulate extreme lockdown scenarios as shown in Table 10.

Table 10. Testing extreme lockdown scenarios.

Figure	Scenario	Immunity Length	Recovered Immunity Protection	Vaccine Protection 1st Dose	Vaccine Protection 2nd Dose	Delay before Lockdown	Lockdown Daily Case Threshold	Lockdown%
9a	Long delay & high case threshold	8/12 months	70%	PB/AZ 65%	PB/AZ 70%	21 days	100,000	20% addition
9b	Severe lockdown	8/12 months	70%	PB/AZ 65%	PB/AZ 70%	7 days	50,000	40% addition

The extreme effects of a high threshold of 100,000 cases and a 21-day delay before lockdown initiation were projected in Figure 9a; for the 8-month immunity base case, the case threshold is reached in August 2021 and lockdown is initiated in early September 2021, continuing for 2 months with daily known cases peaking at 250,000. For the 12-month immunity scenario, a shorter lockdown starting in September is required, and daily cases peak at 160,000. Figure 9b projects the effect of a 40% lockdown increase rather than the 20% used in other scenarios and shows how, for the 8-month immunity base case, reduced transmission opportunity lowers daily cases from a peak of 107,000 to below the 50,000-case threshold, requiring another lockdown phase in late 2021 to reduce case numbers again. The 12-month immunity scenario only requires one lockdown to control case numbers as ongoing vaccinations continue to reduce the susceptible percentage.

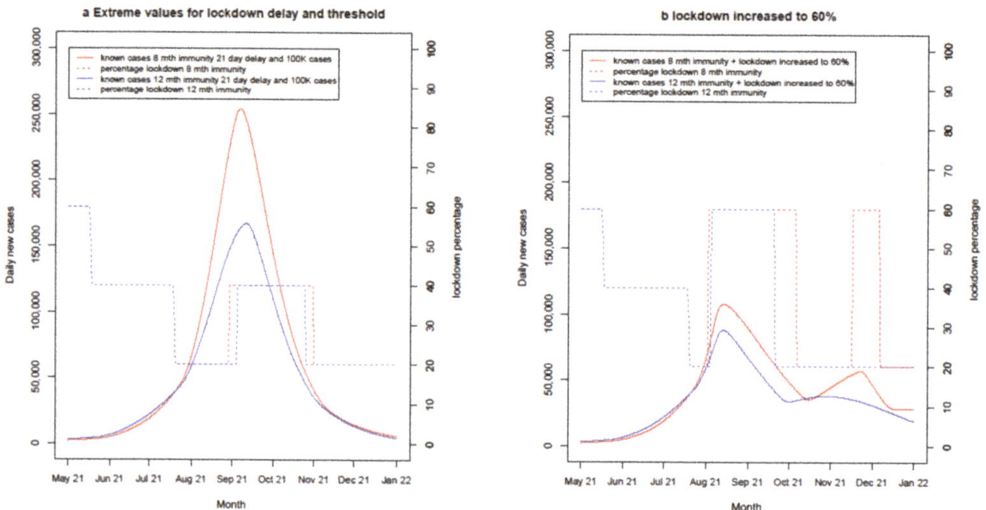

Figure 9. Extreme simulations for lockdowns.

4.2.6. Change in Susceptible Percentage

In February 2021, 100% of the UK population was susceptible to infection with COVID-19. The susceptible percentage dropped as people became immune either through infection or vaccination. The movement of the susceptible percentage is illustrated in Figure 10 for immunity length variation scenarios, with and without new lockdown interventions after June 2021, as shown in Table 11.

Table 11. Susceptible percentage illustrations.

Figure	Scenario	Immunity Length	Recovered Immunity Protection	Vaccine Protection 1st Dose	Vaccine Protection 2nd Dose	Delay before Lockdown	Lockdown Daily Case Threshold	Lockdown%
10a	Immunity variations	5/8/12 months	70%	PB/AZ 65%	PB/AZ 70%	-	-	-
10b	Immunity variations with lockdown intervention	5/8/12 months	70%	PB/AZ 65%	PB/AZ 70%	7 days	50,000	20% addition

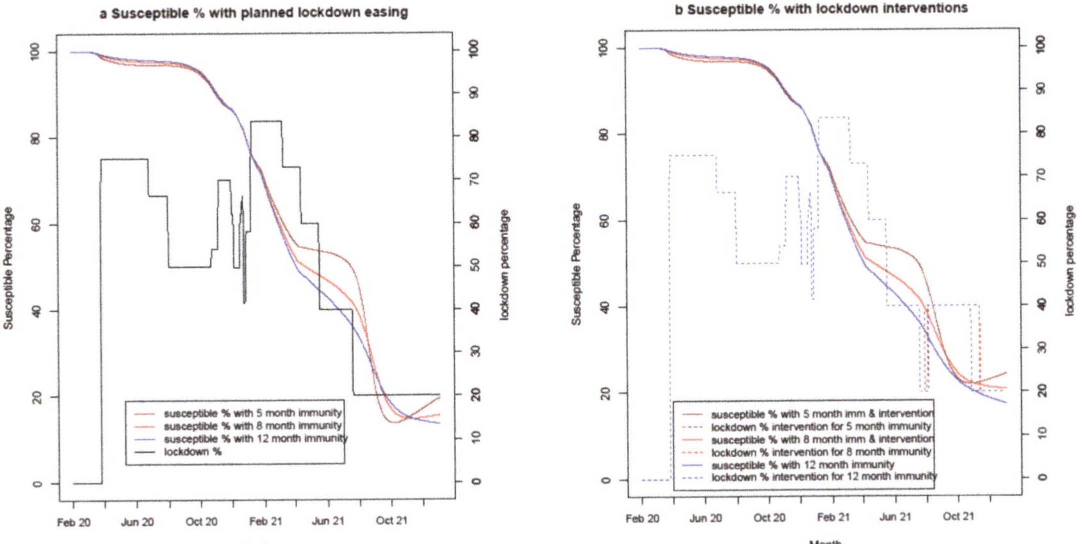

Figure 10. Susceptible population percentage with differing immunity and interventions.

Figure 10 shows the susceptible percentage reducing as the pandemic progresses. The steeper downward slopes correlate with periods of higher infection rates during which more people acquire recovered immunity. In Figure 10a, for the 5-month immunity scenario, the susceptible percentage drops slowly through April to July 2021 as increasing numbers are vaccinated. It then falls steeply to 13% because the infection surge, which is seen in Figure 3, generates recovered immunity before increasing in September 2021 as this immunity erodes. The 8-month and 12-month immunity scenarios follow a similar pattern but with less pronounced slope changes.

Figure 10b shows the susceptible percentages for the three immunity scenarios with lockdown interventions implemented. For all scenarios, lockdowns as illustrated in Figure 6 are required to reduce daily known cases below 50,000. These have the effect of slowing the susceptible percentage reduction by reducing case numbers and hence generating less recovered immunity.

5. Discussion

5.1. Implications of Findings

The UK Government's approach to the COVID-19 pandemic in the UK, though initially hesitant, turned around in early 2021 when strong lockdown measures were put in place and an ambitious vaccination programme was commenced. The UK's aggressive pursuit of vaccination is paying off, with half the population fully vaccinated at the beginning of July 2021. Were it not for the emergence of the Delta variant, assuming that immunity gained from either infection or vaccination lasts at least 8 months, the UK would be assured that it could lift restrictions and keep COVID-19 case numbers at a low level throughout the remainder of 2021. However, sharply rising case numbers in July 2021 are changing the landscape, with health workers once again fearful of being overwhelmed by COVID-19 cases [48]. The vaccination programme has reduced both the transmission and severity of the disease, meaning that hospitalisation and death rates will be greatly reduced, but with half the population still unvaccinated or incompletely vaccinated, and the scenarios projecting hundreds of thousands of daily cases, daily deaths are likely to reach into the hundreds [36] without containment measures.

The most significant influencer of ongoing infection rates, other than the emergence of another more infectious variant, is likely to be the length of protection conferred by vaccinated and recovered immunity. Immunity length is a significant unknown, which will only become clearer as results from longitudinal studies on vaccinated and recovered individuals emerge. The modelling used by the UK Government's SAGE advisory group [13] specifically excludes waning immunity and the future emergence of variants, so these are significant gaps. There are no tools to predict the profile of future variants but further research to understand immunity length, particularly vaccinated immunity, which has a more significant influence in the UK than recovered immunity, is critical for informing policy and for reducing the uncertainty surrounding the various scenarios.

As cases surge, the vaccinated sector of the population will be protected from serious illness and death but vaccination status in the UK is uneven, with lower uptake amongst disadvantaged groups and ethnic minorities, leaving these groups vulnerable. The unvaccinated population will only be effectively protected through herd immunity, which research indicates will be reached with a susceptible percentage of 30% or less [49,50]. The limits on the percentage of the population able to be vaccinated will become the main constraint to achieving herd immunity. About 22% of the UK population are not currently eligible for vaccination (21% under 18, 0.7% pregnant), which means that 90% of eligible adults need to be vaccinated to achieve a 70% total. With the highest infection prevalence in teenagers and 20–24-year-olds [47], extending vaccinations to children is a logical next step to increasing herd immunity, and further research and trials on the safety and efficacy of vaccines for children and pregnant women are required to inform policy. Continuing education and reassurance for the vaccine-hesitant sector of the population is also required to address resistance. It seems likely that for herd immunity to be maintained, regular booster doses of COVID-19 vaccinations will be needed; the practice of immunizing newly eligible people will be insufficient to control the spread of the virus.

Cases are likely to shift from known to unknown because of the reduction in infection severity post-vaccination. As nothing other than lockdown appears to work when there are many unknown cases, a capability which maintains or improves the proportion of known cases is important. The potential for more unknown cases, explored in Section 4.2.3, is a concern and strengthening policies which encourage routine testing mitigates against the growing unknown proportion, and thus the unseen burden of disease. The projections for known cases in Figures 3 and 4 are based on the known proportion remaining at 50%, which is why they are so high in some scenarios.

The current Government policy of 7 day's warning of a change in lockdown status seems a reasonable balance between people's need for notice and the infection growth which takes place in those 7 days, although there is a case for reducing notice to curb growth. Any argument for a low lockdown case threshold to curb growth has been overtaken by events in July 2021, with over 50,000 daily cases being reported. The load on the health services will be a critical consideration in decisions about further restrictions; modelling that is outside the scope of this article.

5.2. Modelling Discussion

The UK Government's SAGE advisory group uses three models from the Imperial College London, Warwick University and the London School of Medicine and Tropical Hygiene groups [13]. The assumptions used by the models are documented, but the public cannot easily see or understand the models or the process by which the results are obtained. This generates mistrust and skepticism, especially as the incorporation of new factors such as the emergence of the Delta variant cannot be done instantaneously.

This model, whilst it has more limitations than the larger models, has the advantage of being able to be displayed on one page, making it potentially more accessible and transparent. It is an aggregated model, with no split into age bands with their differing profiles and vulnerabilities. It does not account for urban/rural differences or for country differences within the UK. Many aspects of the simulation, for example, vaccine rollout ramp up and the emergence of the Alpha and Delta variants, are simplified. However, it is a useful tool for representing COVID-19 transmission in the UK and can be used to project the effects of policies and interventions across a range of uncertainties.

5.2.1. Uncertainty

The model is based on a set of significant assumptions based on evolving clinical research which suggests a range of scenarios. Of particular importance are:

- length of recovered immunity;
- vaccine efficacy in reducing transmission;
- duration and relative infectiousness of asymptomatic and mildly symptomatic cases;
- ongoing uncertainty on the proportion of unknown cases which continue to drive infections.

The strategies for dealing with uncertainty in COVID-19 modelling proposed by Wang/Flessa [51] have been followed for this modelling exercise. It is evident both from the results and the discussion that changes in key assumptions, including future lockdown percentages, can have significant impacts on the projections in the model. Changes in the vaccine mix may also change the model projections. Every month that the pandemic progresses, new research with a direct bearing on the model assumptions is produced, so there is an opportunity for ongoing refinement.

5.2.2. Confidence in the Results for Given Assumptions

An important decision in the modelling process is which values to fix as constants and which to determine through a 'try for fit' calibration process. If one attempts to vary all of the assumed values in the model, there are too many degrees of freedom to be able to obtain meaningful results. It is certainly possible to obtain similar results with different parameter values, in line with the concept of equifiniality, which demonstrates that different sets of parameters can lead to the same or similar results [52]. There is a balance between fixing assumptions to reduce the number of values in play, enabling a meaningful optimization process to be run, and choosing to fix assumptions which are not certain enough, introducing error into the model. The method used in this exercise, which relies on fixing values which have research backing and calibrating the other values against historical data through a curve-fitting exercise, has introduced a level of rigour to the process.

5.2.3. Comparison with Other Models

A significant difference between this model and many other models produced is the inclusion of loss of immunity. Most of the earlier COVID-19 models excluded loss of immunity, although Struben recognises it as a factor which will need to be considered as the pandemic evolves [53]. One other UK-specific exception is the 'Testing and Tracking in the UK' study from the Wellcome Foundation [54], which concludes that the emergence of a new wave of infection depends on the rate at which immunity is lost. This model supports this finding.

A number of studies investigate the difficult issue of true population infection rates for COVID-19 and the high proportion of unknown infections. The ongoing model comparison reporting from the 'Our World in Data' project [2] lists two well-known models from Professor Neil Ferguson's team at the Imperial College London (ICL) and from the Institute for Health Metrics and Evaluation (IHME), which track the estimated total COVID-19 infections against reported infections for many countries. The ICL model shows, after the 'first wave', when testing was immature, total UK cases varied between four and six times the number of known cases, only reducing to roughly double the known cases in late March 2021. The IHME model is more optimistic, showing the total UK cases as no more than double the number of known cases after the first peak and showing no unknown cases in the UK in late March 2021. This model is more aligned with the ICL model, and we believe its findings to be more plausible on the basis that not all infections will be reported for various reasons including asymptomatic or mild infection. Backcasting studies also support estimates in line with the ICL model [55,56]. None of the models or studies project forwards, so forecasting the known proportions at the current level seems to be the only reasonable option despite the large peaks which are projected.

5.2.4. Generalisation

Finally, whilst this model was built for the UK, the only thing which makes it country-specific is the calibration of the parameters and the lockdown profile. It may not be suitable for countries with lower case rates, where factors such as the efficiency of contact tracing have more influence, but otherwise, it is structurally generic and could be adapted for other countries or regions. Whilst decisions in managing this pandemic cannot be based on modelling alone, the predictive power of dynamic modelling can serve as a powerful tool to inform policies and intervention decisions. Never has modelling been more important in the field of public health.

6. Conclusions

Whilst there continues to be considerable uncertainty surrounding the progression of the COVID-19 pandemic in the UK, this modelling exercise identifies the key factors generating this uncertainty and projects the results of lockdown changes under a variety of scenarios. UK policy makers set a reasonable course to enable the countries to exit from lockdown, but the infection surge resulting from the emergence of the Delta variant has given yet another challenge which can only be addressed by an ongoing focus on vaccination and potentially by further social restrictions.

The model, whilst by no means perfect, is useful for projection purposes, and its simplicity and transparency are meant to provide further insight to the modelling and analysis process to both policy makers and the general public. As with any model, the assumptions behind it are critical to its accuracy. New COVID-19 research is being published all the time, and the model can continue to be refined and updated as both the research and policy evolves and more historical data is produced.

Supplementary Materials: The following are available online at https://www.mdpi.com/article/10.3390/systems9030060/s1.

Author Contributions: This manuscript was created by C.B. and reviewed and enhanced by M.P. All authors have read and agreed to the published version of the manuscript.

Funding: This research received no external funding.

Institutional Review Board Statement: Not applicable.

Informed Consent Statement: Not applicable.

Data Availability Statement: All data used in this study can be found at www.ourworldindata.org or in indicated references.

Acknowledgments: The authors wish to acknowledge Robert Eberlein at isee systems for the original sample model which provided the starting point for this modelling activity. This research was partially financially supported by a private donation to the Millennium Institute. The donor had no role in study design, data collection and analysis, decision to publish, or preparation of the manuscript.

Conflicts of Interest: The authors declare no conflict of interest.

References

1. Godlee, F.; Silberner, J. The BMJ interview: Anthony Fauci on covid-19. *BMJ* **2020**, *370*, m3703. [CrossRef]
2. How Epidemiological Models of COVID-19 Help Us Estimate the True Number of Infections. Available online: https://ourworldindata.org/covid-models (accessed on 30 June 2021).
3. Rahmandad, H.; Lim, T.Y.; Sterman, J. Behavioral dynamics of COVID-19: Estimating under-reporting, multiple waves, and adherence fatigue across 91 nations. *medRxiv* **2020**. [CrossRef]
4. Ward, H.; Cooke, G.; Atchison, C.; Whitaker, M.; Elliott, J.; Moshe, M. Spiral: Antibody prevalence for SARS-CoV-2 in England following first peak of the pandemic: REACT2 study in 100,000 adults. *bioRxiv* **2020**. [CrossRef]

5. COVID-19 Pandemic Planning Scenarios. Available online: https://www.cdc.gov/coronavirus/2019-ncov/hcp/planning-scenarios.html (accessed on 30 March 2021).

6. We Scientists Said Lock Down. But UK Politicians Refused to Listen. Available online: https://www.theguardian.com/commentisfree/2020/apr/15/uk-government-coronavirus-science-who-advice (accessed on 30 March 2021).

7. Hancock, M. An Update on the Coronavirus Vaccine, 2 December 2020. Available online: https://www.gov.uk/government/speeches/an-update-on-the-coronavirus-vaccine-2-december-2020 (accessed on 30 March 2021).

8. Covid-19: Oxford-AstraZeneca Vaccine Approved for Use in UK. Available online: https://www.bbc.com/news/health-55280671 (accessed on 30 March 2021).

9. COVID-19 Response—Spring 2021. Available online: https://www.gov.uk/government/publications/covid-19-response-spring-2021 (accessed on 30 March 2021).

10. Imperial College COVID-19 Reports. Available online: https://mrc-ide.github.io/global-lmic-reports/ (accessed on 30 June 2021).

11. Holmdahl, I.; Buckee, C. Wrong but Useful—What Covid-19 Epidemiologic Models Can and Cannot Tell Us. *N. Engl. J. Med.* **2020**, *383*, 303–305. [CrossRef]

12. IHME COVID-19 Projections. Available online: https://covid19.healthdata.org/global?view=cumulative-deaths&tab=trend (accessed on 30 June 2021).

13. SPI-M-O: Summary of Modelling on Scenarios for Easing Restrictions. Available online: https://assets.publishing.service.gov.uk/government/uploads/system/uploads/attachment_data/file/975909/S1182_SPI-M-O_Summary_of_modelling_of_easing_roadmap_step_2_restrictions.pdf (accessed on 30 April 2021).

14. Cohen, J.I.; Burbelo, P.D. Reinfection with SARS-CoV-2: Implications for Vaccines. *Clin. Infect. Dis.* **2020**, cia1866. [CrossRef]

15. Dan, J.M.; Mateus, J.; Kato, Y.; Hastie, K.M.; Yu, E.D.; Faliti, C.E.; Grifoni, A.; Ramirez, S.I.; Haupt, S.; Fraizer, A.; et al. Immunological memory to SARS-CoV-2 assessed for up to eight months after infection. *bioRxiv* **2020**. [CrossRef]

16. Seow, J.; Graham, C.; Merrick, B.; Acors, S.; Steel, K.J.A.; Hemmings, O.; O'Bryne, A.; Kouphou, N.; Pickering, S.; Galao, R.P.; et al. Longitudinal evaluation and decline of antibody responses in SARS-CoV-2 infection. *medRxiv* **2020**. [CrossRef]

17. Wajnberg, A.; Amanat, F.; Firpo, A.; Altman, D.R.; Bailey, M.J.; Mansour, M.; McMahon, M.; Meade, P.; Mendu, D.R.; Muellers, K.; et al. SARS-CoV-2 infection induces robust, neutralizing antibody responses that are stable for at least three months. *medRxiv* **2020**. [CrossRef]

18. Mumoli, N.; Vitale, J.; Mazzone, A. Clinical immunity in discharged medical patients with COVID-19. *Int. J. Infect. Dis.* **2020**, *99*, 229–230. [CrossRef] [PubMed]

19. Mahase, E. Covid-19: Past infection provides 83% protection for five months but may not stop transmission, study finds. *BMJ* **2021**, *372*, n124. [CrossRef]

20. Pritchard, E.; Matthews, P.C.; Stoesser, N.; Eyre, D.W.; Gethings, O.; Vihta, K.-D.; Jones, J.; House, T.; VanSteenHouse, H.; Bell, I.; et al. Impact of vaccination on SARS-CoV-2 cases in the community: A population-based study using the UK's COVID-19 Infection Survey. *medRxiv* **2021**. [CrossRef]

21. Hodgson, S.H.; Mansatta, K.; Mallett, G.; Harris, V.; Emary, K.R.W.; Pollard, A.J. What defines an efficacious COVID-19 vaccine? A review of the challenges assessing the clinical efficacy of vaccines against SARS-CoV-2. *Lancet Infect. Dis.* **2021**, *21*, e26–e35. [CrossRef]

22. Rivett, L.; Sridhar, S.; Sparkes, D.; Routledge, M.; Jones, N.K.; Forrest, S.; Young, J.; Pereira-Dias, J.; Hamilton, W.L.; Ferris, M.; et al. Screening of healthcare workers for SARS-CoV-2 highlights the role of asymptomatic carriage in COVID-19 transmission. *eLife* **2020**, *9*, e58728. [CrossRef] [PubMed]

23. Lee, S.; Meyler, P.; Mozel, M.; Tauh, T.; Merchant, R. Asymptomatic carriage and transmission of SARS-CoV-2: What do we know? *Can. J. Anesth.* **2020**, *67*, 1424–1430. [CrossRef]

24. Altmann, D.M.; Douek, D.C.; Boyton, R.J. What policy makers need to know about COVID-19 protective immunity. *Lancet* **2020**, *395*, 1527–1529. [CrossRef]

25. Johansson, M.A.; Quandelacy, T.M.; Kada, S.; Prasad, P.V.; Steele, M.; Brooks, J.T.; Slayton, R.B.; Biggerstaff, M.; Butler, J.C. SARS-CoV-2 Transmission from People without COVID-19 Symptoms. *JAMA Netw. Open* **2021**, *4*, e2035057. [CrossRef] [PubMed]

26. Pollock, A.M.; Lancaster, J. Asymptomatic transmission of Covid-19. *BMJ* **2020**, *371*, m4851. [CrossRef]

27. Byambasuren, O.; Cardona, M.; Bell, K.; Clark, J.; McLaws, M.-L.; Glasziou, P. Estimating the extent of asymptomatic COVID-19 and its potential for community transmission: Systematic review and meta-analysis. *Off. J. Assoc. Med. Microbiol. Infect. Dis. Can.* **2020**, *5*, 223–234. [CrossRef]

28. Lau, H.; Khosrawipour, T.; Kocbach, P.; Ichii, H.; Bania, J.; Khosrawipour, V. Evaluating the massive underreporting and undertesting of COVID-19 cases in multiple global epicenters. *Pulmonology* **2021**, *27*, 110. [CrossRef] [PubMed]

29. Kermack, W.; Mckendrick, A.G. A Contribution to the Mathematical Theory of Epidemics. *Proc. R. Soc. A Math. Phys. Eng. Ser.* **1927**, *115*, 700–721.

30. Ward, H.; Graham, C.; Atchison, C.; Whitaker, M.; Elliot, J.; Moshe, M.; Brown, J.C.; Flower, B.; Daunt, A.; Ainslie, K.; et al. Declining prevalence of antibody positivity to SARS-CoV-2: A community study of 365,000 adults. *bioRxiv* **2020**. [CrossRef]

31. Volz, E.; Mishra, S.; Chand, M.; Barrett, J.; Johnson, R.; Geidelberg, L.; Hinsley, W.R.; Laydon, D.J.; Dabrera, G.; O'Toole, A.; et al. Transmission of SARS-CoV-2 Lineage B.1.1.7 in England: Insights from linking epidemiological and genetic data. *medRxiv* **2020**. [CrossRef]

32. Graham, M.S.; Sudre, C.H.; May, A.; Antonelli, M.; Murray, B.; Varsavsky, T.; Kläser, K.; Canas, L.S.; Molteni, E.; Modat, M.; et al. Changes in symptomatology, reinfection, and transmissibility associated with the SARS-CoV-2 variant B.1.1.7: An ecological study. *Lancet Public Health* **2021**, *6*, 335–380. [CrossRef]

33. Moss, R.; Wood, J.; Brown, D.; Shearer, F.; Black, A.J.; Cheng, A.C.; McCaw, J.M.; McVernon, J. Modelling the impact of COVID-19 in Australia to inform transmission reducing measures and health system preparedness. *medRxiv* **2020**. [CrossRef]

34. Byrne, A.W.; McEvoy, D.; Collins, A.B.; Hunt, K.; Casey, M.; Barber, A.; Butler, F.; Griffin, J.; Lane, E.A.; McAloon, C.; et al. Inferred duration of infectious period of SARS-CoV-2: Rapid scoping review and analysis of available evidence for asymptomatic and symptomatic COVID-19 cases. *BMJ Open* **2020**, *10*, e039856. [CrossRef] [PubMed]

35. UK Government. Joint Committee on Vaccination and Immunisation: Advice on Priority Groups for COVID-19 Vaccination, 30 December 2020. Available online: https://www.gov.uk/government/publications/priority-groups-for-coronavirus-covid-19-vaccination-advice-from-the-jcvi-30-december-2020/joint-committee-on-vaccination-and-immunisation-advice-on-priority-groups-for-covid-19-vaccination-30-december-2020 (accessed on 30 March 2021).

36. Ritchie, H.; Ortiz-Ospina, E.; Beltekian, D.; Mathieu, E.; Hasell, J.; Macdonald, B.; Giattino, C.; Appel, C.; Rodés-Guirao, L.; Roser, M. Coronavirus Pandemic (COVID-19). Available online: https://ourworldindata.org/coronavirus (accessed on 12 July 2021).

37. Robertson, E.; Reeve, K.S.; Niedzwiedz, C.L.; Moore, J.; Blake, M.; Green, M.; Katikireddi, S.V.; Benzeval, M.J. Predictors of COVID-19 vaccine hesitancy in the UK Household Longitudinal Study. *Brain Behav. Immun.* **2021**, *94*, 41–50. [CrossRef]

38. Liu, Y.; Tang, J.W.; Lam, T.T.Y. Transmission dynamics of the COVID-19 epidemic in England. *Int. J. Infect. Dis.* **2021**, *104*, 132–138. [CrossRef]

39. Noh, J.Y.; Yoon, J.G.; Seong, H.; Choi, W.S.; Sohn, J.W.; Cheong, H.J.; Kim, W.J.; Song, J.Y. Asymptomatic infection and atypical manifestations of COVID-19: Comparison of viral shedding duration. *J. Infect.* **2020**, *81*, 816–846. [CrossRef]

40. Zhou, R.; Li, F.; Chen, F.; Liu, H.; Zheng, J.; Lei, C.; Wu, X. Viral dynamics in asymptomatic patients with COVID-19. *Int. J. Infect. Dis.* **2020**, *96*, 288–290. [CrossRef] [PubMed]

41. Increased Household Transmission Of COVID-19 Cases Associated with SARS-Cov-2 Variant of Concern B.1.617.2: A National Case-Control Study. Available online: https://khub.net/documents/135939561/405676950/Increased+Household+Transmission+of+COVID-19+Cases+-+national+case+study.pdf/7f7764fb-ecb0-da31-77b3-b1a8ef7be9aa (accessed on 31 March 2021).

42. Atalan, A. Is the lockdown important to prevent the COVID-9 pandemic? Effects on psychology, environment and economy-perspective. *Ann. Med. Surg.* **2020**, *56*, 38–42. [CrossRef] [PubMed]

43. Prime Minister Sets Out Roadmap to Cautiously Ease Lockdown Restrictions. Available online: https://www.gov.uk/government/news/prime-minister-sets-out-roadmap-to-cautiously-ease-lockdown-restrictions (accessed on 30 April 2021).

44. Wright, L.; Fancourt, D. Do predictors of adherence to pandemic guidelines change over time? A panel study of 21,000 UK adults during the COVID-19 pandemic. *medRxiv* **2020**. [CrossRef]

45. Sanche, S.; Lin, Y.T.; Xu, C.; Romero-Severson, E.; Hengartner, N.; Ke, R. High Contagiousness and Rapid Spread of Severe Acute Respiratory Syndrome Coronavirus 2. *Emerg. Infect. Dis.* **2020**, *26*, 1470–1477. [CrossRef]

46. Subsidizing the Spread of COVID19: Evidence from the UK's Eat-Out-to-Help-Out Scheme. Available online: https://ideas.repec.org/p/wrk/warwec/1310.html (accessed on 31 January 2021).

47. Prime Minister Confirms Move to Step 4. Available online: https://www.gov.uk/government/news/prime-minister-confirms-move-to-step-4 (accessed on 12 July 2021).

48. Rise in Covid Cases Will Put Intense Pressure on NHS, Bosses Warn. Available online: https://www.theguardian.com/world/2021/jul/12/rise-in-covid-cases-will-put-intense-pressure-on-nhs-bosses-warn (accessed on 14 July 2021).

49. Britton, T.; Ball, F.; Trapman, P. A mathematical model reveals the influence of population heterogeneity on herd immunity to SARS-CoV-2. *Science* **2020**, *369*, 846–849. [CrossRef] [PubMed]

50. Fontanet, A.; Cauchemez, S. COVID-19 herd immunity: Where are we? *Nat. Rev. Immunol.* **2020**, *20*, 583–584. [CrossRef]

51. Wang, M.; Flessa, S. Modelling Covid-19 under uncertainty: What can we expect? *Eur. J. Health Econ.* **2020**, *21*, 665–668. [CrossRef] [PubMed]

52. Beven, K. A manifesto for the equifinality thesis. *J. Hydrol.* **2006**, *320*, 18–36. [CrossRef]

53. Struben, J. The coronavirus disease (COVID-19) pandemic: Simulation-based assessment of outbreak responses and postpeak strategies. *Syst. Dyn. Rev.* **2020**, *36*, 247–293. [CrossRef] [PubMed]

54. Friston, K.J.; Parr, T.; Zeidman, P.; Razi, A.; Flandin, G.; Daunizeau, J.; Hulme, O.J.; Biling, A.J.; Litvak, J.; Price, C.J.; et al. Tracking and tracing in the UK: A dynamic causal modelling study. *Wellcome Open Res.* **2021**, *5*, 144. [CrossRef]

55. Phipps, S.J.; Grafton, R.Q.; Kompas, T. Robust estimates of the true (population) infection rate for COVID-19: A backcasting approach. *R. Soc. Open Sci.* **2020**, *7*, 200909. [CrossRef]

56. Böhning, D.; Rocchetti, I.; Maruotti, A.; Holling, H. Estimating the undetected infections in the Covid-19 outbreak by harnessing capture–recapture methods. *Int. J. Infect. Dis.* **2020**, *97*, 197–201. [CrossRef]

Article

Implementation of an Expanded Decision-Making Technique to Comment on Sweden Readiness for Digital Tourism

Shahryar Sorooshian

Department of Business Administration, University of Gothenburg, 41124 Gothenburg, Sweden; sorooshian@gmail.com

Abstract: Tourism provides many advantages for Sweden and the whole world, as well as its travelers. Since almost all types of tourism are currently in crisis as a result of the current COVID-19 pandemic, information and communication technology is expected to play a role, not only during the crisis but also in the post-COVID-19 era. Thus, with no expectations from types of tourism, Sweden needs to broaden its digital tours. As a result, this letter aims to classify the transition readiness of industry clusters for this digitalization move. An extended version of the TOPSIS technique was formulated and validated, plus a new framework for measuring digitalization readiness for this purpose. Lastly, analysis of the collected data proves that business tourism could lead the change, though adventure and rural tourism are at the farthest point from being considered ready to change.

Keywords: tour and traveling; digitalization shift; change readiness; expanded TOPSIS; COVID-19

Citation: Sorooshian, S. Implementation of an Expanded Decision-Making Technique to Comment on Sweden Readiness for Digital Tourism. *Systems* **2021**, *9*, 50. https://doi.org/10.3390/systems9030050

Academic Editors: Oz Sahin, Russell Richards and Jan Kwakkel

Received: 17 May 2021
Accepted: 1 July 2021
Published: 5 July 2021

Publisher's Note: MDPI stays neutral with regard to jurisdictional claims in published maps and institutional affiliations.

1. Introduction

The tourism industry encompasses a broad variety of events, as tourism is described as persons traveling to and staying in places outside their typical environment for a maximum of a year for business, leisure, or any other dedications [1]. Scholars [2] believe that tourism plays a crucial role in the growth and development of all countries. Any crisis for tourism could be a challenge for many subdivisions, as in recent decades, tourism has stretched into various types [3,4]: adventure tourism, urban tourism, cultural tourism, event tourism, etcetera. In 2019, approximately 1.5 billion international tourist arrivals were estimated worldwide, and prior to the 2020 pandemic, international travel was forecast to expand more than three percent per year [5].

The World Health Organization (WHO) announced a worldwide pandemic in March 2020: the COVID-19 pandemic, which was a disease caused by the SARS-CoV-2 coronavirus. The COVID-19 pandemic is still a challenge in 2021 for the whole world. More or less, people are in mandatory quarantine or quarantine of their own volition and travels are minimized due to the pandemic [6]. Accordingly, the tourism industry is facing a crisis due to this virus. It is a serious issue, as Peceny et al. [7] said that even a slight change in this industry has a massive impact on all of society. Although the tourism industry has experienced different crises, the impact of the current crisis is more shocking than any earlier ones, at least from an economical perspective [8]. Thus, many professionals, including Higgins-Desbiolles [9] and Gretzel et al. [8], call for an urgent solution for the industry to handle and recover from this crisis. However, which types of the tourism industry should be targeted for urgent intervention need to be assessed, as well as which types are capable of better adapting to the circumstances.

Scholars argue that tourism is not only generating financial growth and job opportunities, but also significantly contributes to quality of life [2]. However, this pandemic lockdown has caused a negative impact on people's daily lives and several reports have recently alerted us about the mental health burden of this pandemic (i.e., [10–13]). Due to this pandemic, a lot of people are suffering from heightened mental health problems, such as depression, anxiety, and sadness, which have emerged as significant public health challenges. These can also lead to severe behavioral and physical health issues with serious effects, with both social and personal costs [10]. Therefore, studies to mitigate this mental health burden are called for by many scientists [14,15]. It would be interesting to see how reactivation of the tourism industry can play a role in solving this mental health burden due to the lockdowns. The solution could be transportation-free tourism; however, this has yet to be thoroughly researched.

Clearly, another impact of this pandemic is the extensive rise in the use of ICT [16,17]. Garfin [16] said, while considering possible negative consequences, that a thoughtful approach to using ICT can be effective and necessary for coping during the COVID-19 pandemic and as societies move into a new future. COVID-19 is a psychological framing of what might result in post-pandemic tourism behavior [18]. Garfin [16] believed that the ongoing COVID-19 pandemic provides opportunities to investigate core values for expanding the conscientious use of technology to mitigate the negative impact of stress and improve people's lives. This pandemic heightens the significant importance of ICT, even though this technology had influenced different aspects of people's daily life for a long time [1,19]. However, among the advantages of ICT implementation is permitting processes to be accessible with subordinate cost and additional efficiency [1].

Additionally, Chamarro [20] said it is very clear that people's lives after the COVID-19 crisis will be marked by the experience of intensive use of ICT during the pandemic. In tourism there is evidence for the successful implementation of ICT [21]; therefore, it is predictable that digitalization will remain in tourism, even after COVID-19, as a new normal [8,22].

Not only now, but even long before COVID-19, the ICT industry began to collaborate with tourism. The phenomenon of digital tours has arisen from the integration of information technology and tourism [5]. Digital tours cannot be a negative trend because they are expected to decrease some of the industry's severe consequences. Traditional tourism contributes significantly to the rising levels of air pollution [23], and the negative result on the host nation includes noise, overcrowding, and pollution with leftovers [24], as well as the probability of losing cultural values and authenticity, as noted by Ogarlaci and Tonea [23]. In addition, traditional travelers are concerned about political risks such as political instability and terrorism, as well as other hazards for travelers due to natural catastrophes, and a lack of healthcare and clean food or water [25]. The entire list of unfavorable industry outcomes is lengthier, and identifying them requires an individual extensive literature study, but the positive outcomes are also numerous.

Nonetheless, for good or bad, the COVID-19 pandemic has rapidly catapulted ICT to the forefront of people's lives. Now it has significantly exacerbated long-foreseen patterns; it has rapidly pushed a lot of industries that have been able to operate remotely. In brief, ICT has made a major impact on the travel industry [26] and now the industry should be based on this consumer-centric technology in order to satisfy the emerging experienced customers [27]. Hence, it is expected that a positive trend of interest in digital tours in the post-COVID era will be seen.

Not only is there a digitalization push from the COVID-19 pandemic, but also from a different perspective, the Fourth Industrial Revolution (also sometimes known as Industry 4.0) in recent years has rapidly been upsetting industries, including the tourism industry [28]. Tourism is greatly involved in Industry 4.0 digital transformation [29]. Tourism 4.0, as defined by Peceny et al. [7], involves reducing the harmful effects of tourism (e.g., tourism's carbon footprint) and simultaneously improving it through the merging of ICT with the tourism experience. This has turned out to be key to resilience in tourism [8].

Considering the fact that digital tours are an essential supplement for the industry not only during the COVID-19 crisis but also for future tourism, the digitalization readiness of the industry is the key foundation of tourism success. In theory, a crucial step to understanding the capacity to launch and accept a change in ways that provide value, limit risk, and sustain performance is referred to as readiness measurement [30]. Despite the importance of the area, very few studies contribute to this important field, especially when it comes to the different types of tourism in the industry. Although there is attractiveness and pushes for virtual and digital travels, stakeholders' readiness (service supply and travelers' demand) to transition to the modern industry is critical [27], as is technology usability [31]. With regard to the digitalization of tours as a change in an industry system, the author uses a framework with three readiness metrics out of the system theory perspective [32], with input (supply) response, output (demand) response, and process (technology) readiness, as shown in Figure 1, due to the gap in the literature. In a separate article [33], the author gives insight into these readiness metrics.

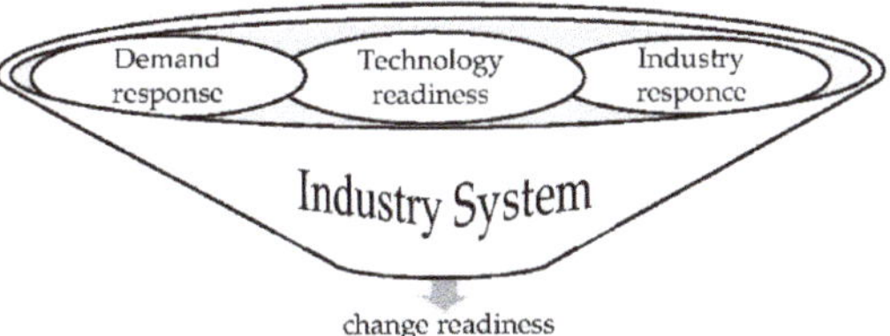

Figure 1. Change readiness from a system approach.

This study focuses on tourism digitalization readiness; however, it is willing to address tourism in Sweden in order to reach a decision on the specific goal of this study. The tourism industry in Sweden has a considerable turnover and plays a noteworthy role in several respects [34]. Sweden is among the top digitalized EU (European Union) economies [35], though there is not enough research focusing on the tourism industry in Sweden. Due to the different characteristics and approaches of societies, studies from other countries may not be fully applicable here. More studies on the tourism industry of Sweden are needed [36], so Sweden's tourism industry is targeted as the scope of this research. Different types of tourism in the industry, with different levels of digitalization readiness, are active in the country. Hence, comparing the readiness for digital tourism could provide a better understanding of capabilities, available benchmarks, and digitalization implementation experiences for Swedish tourism policymaking. In a few words, the main goal of this article is to compare different types of tourism in Sweden based on their readiness for a digitalization shift in order to answer the question, "Which types of tourism in Sweden are more (or less) prepared for the digitalization switchover?"

2. Method

Three criteria were defined to measure the change readiness of the industry—demand response, industry response, and technology readiness—hence, multi-criteria decision-making (MCDM) approaches were targeted for this research. Sweden is active in more than one type of tourism; hence, among MCDM approaches, techniques from multiple-attribute decision-making (MADM) are appropriate. The hierarchical structure of this research is constructed in Figure 2.

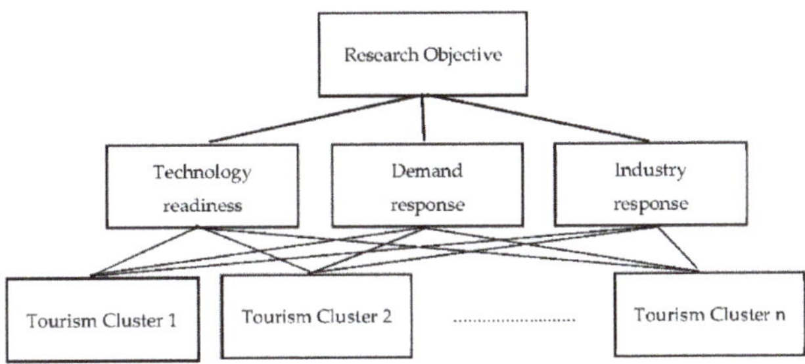

Figure 2. Hierarchical structure of this research.

An ideal tourism cluster that fully satisfies all three readiness measures does not exist practically, so the selected MADM techniques should approximate and list the closest clusters to the ideal. MADM-TOPSIS (techniques for order preference by similarity to an ideal solution) is based on the principle that the listing of the alternatives must be with the concept that priority is given to the option closest to the ideal and the one farthest away from the worst [37,38]. The TOPSIS approach has successfully addressed numerous real-world issues, particularly in recent years, due to its rationality [39]; its accuracy was compared to other MADM techniques and it was recommended [40]. Applied mathematical modeling [38] has communicated the hierarchical structure of TOPSIS per Figure 3. The definition of TOPSIS as ranking the alternatives has received progressive attention from researchers who focus on multiple criteria decision-making approaches [37].

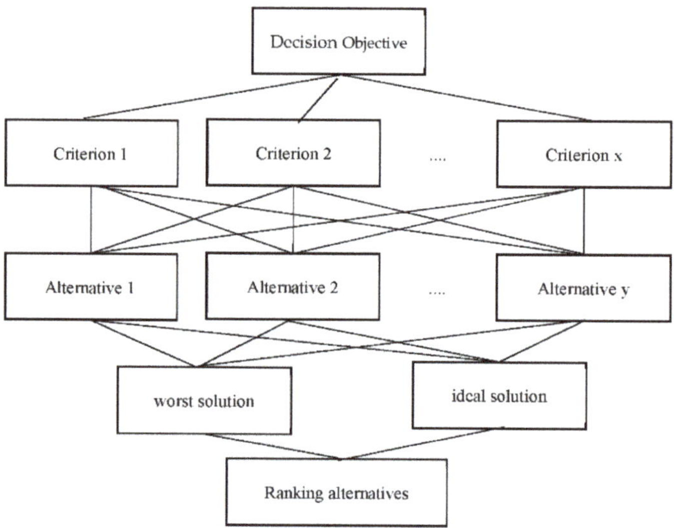

Figure 3. Hierarchical structure of TOPSIS.

There are no constraints reported on the distribution of data, the number of alternatives and criteria, or the sample size of experts in this method. A report on optimizing the use by an expert panel [41] indicated that even a handful of experts in a panel were preferred in several published studies to reach a consensus decision, as the quality of the experts is deemed to be more significant than the size of the panel. A bigger panel may cause too much variety in the feedback and result in a high degree of inconsistency. Hence, the number of experts should be kept to a minimum.

Based on a previous practice [42], the calculation steps of the classic TOPSIS process are listed in Figure 4.

To utilize TOPSIS and due to a lack of literature and the novelty of the COVID-19 situation, primary data were required for this study. For tourism-related data, a panel of experts was invited for group decision-making (GDM), which makes use of its members' varied experiences and interests. Since the scope of this research was defined for tourism in Sweden, the panel of experts was professionals in the field in Sweden who had studied the industry and were aware of existing tourism activities in Sweden.

The expert selection process is important for enhancing the reliability and validity of the research results. Hence, a list of experts was selected based on the number of indexed publications in the past three years in Scopus, by searching the two keywords of "tour *" in title, keywords, and abstract, and "Sweden" in affiliation. A dozen scholars were listed with the highest publications, though after reviewing the scope and title of their publications, the eight most relevant authors were invited for data collection regardless of any conciliation such as academic level, gender, or age.

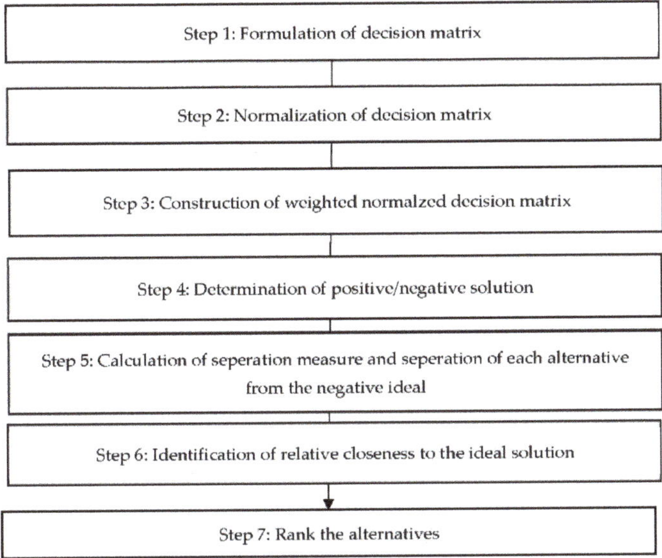

Figure 4. TOPSIS calculations.

Before the data collection in March 2021, in a live online seminar for the pre-study step, a few researchers from the Centre for Tourism (CFT) at Gothenburg University were consulted to comment on the improvement of the prepared data collection tool. In addition, a short follow-up meeting a week later with the seminar chair was organized to review the comments received from the seminar and changes to the data collection instrument. As a result, it was chosen to restrict the spectrum of the study to seven tourism types, and the instrument's framework was designed as shown in screenshot in Figure 5 (for tourism experts). The scales for the answers ranged from −4 to +4, or from "extremely against" to "extremely supportive." The criteria weights were built to accept answers on a 10-point scale, ranging from 10% to 100%.

For technology (ICT) concerns, one expert was invited who had both work-related (nearly 10 years in ICT-related scopes) and related educational backgrounds (with a master's degree in ICT-related fields and a few professional certifications in the area) who also self-reported his awareness of the current ICTs for travel digitalization. Comparably, for ICT-related data collection instruments, there were the same seven types of tourism and similar scales for measuring readiness (technology and user capacity at a fair cost) for digital/virtual travels.

Next, to improve the consensus in the data collection phase, a list of an operational definition of key terms presented to the panels was included in the prepared questionnaires, as shown in Appendix A. Additionally, in the absence of standard terminology in the tourism research literature, it was predicted that supplying this list would yield more reliable results.

Even so, when using MADM techniques such as TOPSIS, it is often assumed that decision-making is conducted with a panel or a task group, and still further work is needed to improve a comprehensive problem-solving technique [39]. Hence, for the analysis of the collected data, classic TOPSIS was not capable to consider inputs from two groups of experts. Sorooshian and Parsia [43] also explained this as one of the constraints of existing MADMs; they suggested a supplementary procedure for solving this issue, called decisions with altered sources of information, which will be included in this study. This procedure adds a few sub-steps to the MADM data entry of that can be summarized as: Step 1, construction of a decision matrix with inputs from the main source of information; Step 2, completion of the decision matrix with inputs from the altered source of information; and Step 3, normalizing the decision matrix.

QUESTION	ANSWER
In *Urban Tourism*, does Demand (**tourists**) support the digitalization of travelings?	Choose an item.
In *Urban Tourism*, does Industry (**service providers**) support the digitalization of travelings?	Choose an item.
In *Cultural Tourism*, does Demand (**tourists**) support the digitalization of travelings?	Choose an item.
In *Cultural Tourism*, does Industry (**service providers**) support the digitalization of travelings?	Choose an item.
In *Rural Tourism*, does Demand (**tourists**) support the digitalization of travelings?	Choose an item.
In *Rural Tourism*, does Industry (**service providers**) support the digitalization of travelings?	Choose an item.
In *Adventure Tourism*, does Demand (**tourists**) support the digitalization of travelings?	Choose an item.
In *Adventure Tourism*, does Industry (**service providers**) support the digitalization of travelings?	Choose an item.
In *Event Tourism*, does Demand (**tourists**) support the digitalization of travelings?	Choose an item.
In *Event Tourism*, does Industry (**service providers**) support the digitalization of travelings?	Choose an item.
In *Business Tourism*, does Demand (**tourists**) support the digitalization of travelings?	Choose an item.
In *Business Tourism*, does Industry (**service providers**) support the digitalization of travelings?	Choose an item.
In *Entertainment Tourism*, does Demand (**tourists**) support the digitalization of travelings?	Choose an item.
In *Entertainment Tourism*, does Industry (**service providers**) support the digitalization of travelings?	Choose an item.

Figure 5. Instrument framework.

Additionally, considering the fact that the research focus of the tourism experts might not cover the whole industry, an add-on consideration of unbalanced expertise was added to the TOPSIS process. For this, experts were asked to refer to the questions asked about each cluster of the tourism industry, and grade their level of expertise. The confidence level for each aspect of the tourism industry was designed to accept scales from 0% to 100%. Sorooshian [44] suggested the application of this confidence/level of expertise through a weighted average of inputs when dealing with group decision-making with a panel of experts with unbalanced expertise.

For the ICT expert, since the needed information was collectible, an assignment was designed and the expert was asked to have the questions but, if needed, answer them after a mini-research (internet search and asking his colleagues) with updated relevant information.

After the above-listed considerations, an expanded TOPSIS, TOPSIS for group decision-making with multiple sources of data through panels of experts with unbalance expertise, was taken into consideration for the data analysis. Figure 6 presents the summary of the steps taken for this study.

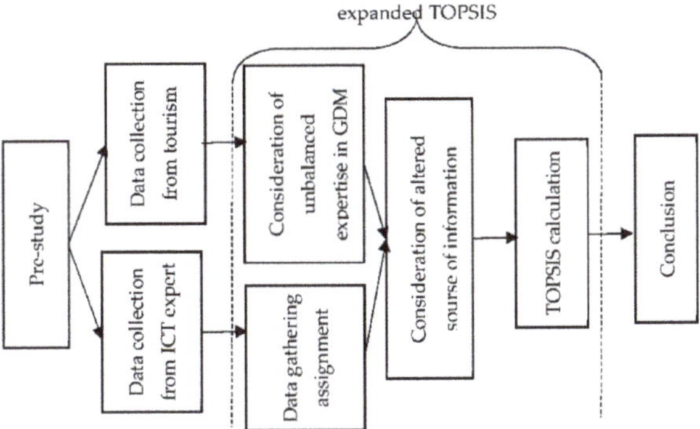

Figure 6. Research flow.

3. Results

After collecting the required data, the calculation steps of the expanded TOPSIS resulted in the following outcomes:

By calculation of the mean, the weights of the decision criteria based on inputs from both panels of experts were calculated: demand response (5.67), industry response (2.67), and technology readiness (7.67).

Part A of Table 1 shows the average inputs from the tourism-panel decision matrix after the consideration of unbalanced expertise for group decision-making. However, part B shows the average input from the ICT-panel decision matrix after the consideration of unbalanced expertise for group decision-making.

Table 1. Decision matrix.

Alternatives	Part A		Part B
	Demand Response	Industry Response	Technology Readiness
Urban tourism	0.43	−1.3	2
Cultural tourism	0.9	0.27	3
Rural tourism	−0.13	−1.17	0
Adventure tourism	−0.67	−0.13	1
Event tourism	1.47	0.53	4
Business tourism	2.03	1.03	4
Entertainment tourism	−0.4	0.27	2

Appendix B shows the output from the application of a web-based software, Decision Radar Ez-TOPSIS (https://decision-radar.com/Topsis.html (accessed on 30 April 2021)), for decision-making with the TOPSIS method.

Finally, Table 2 reports the results of the hierarchy ranking of the tourism clusters starting from the closest to the ideal (fully ready to be digitalized).

Table 2. Results.

Rank	Cluster	Score
1	Business tourism	1.00
2	Event tourism	0.84
3	Cultural tourism	0.67
4	Urban tourism	0.43
5	Entertainment tourism	0.38
6	Adventure tourism	0.24
7	Rural tourism	0.20

Therefore, here in this study, from the analysis of the collected data, business tourism followed by event tourism seems to be more ready than other clusters of the industry for a digitalization shift whenever it is needed. With many of us moving our business work online as a result of COVID-19 and social distancing, the use of video conferencing programs has grown exponentially. Video conferencing promotes long-distance and international connectivity and improves teamwork while minimizing travel costs [45]. There are many video-conferencing programs available, including Skype, Zoom, Facetime, Zoho Meeting, Highfive Meeting, GoToMeeting, Google Hangouts Meet, Slack, Cisco WebEx, and Eyeson, to name a few [46]. For instance, although only 10 million people attended Zoom meetings before COVID-19 became widespread at the end of 2019, consumption had skyrocketed to 300 million by April 2020 [47].

With many cross-country examples, Arshad [46] explained that ICTs have allowed business meetings and events to retain a semblance of normalcy during quarantine, enabling them to transfer their meetings electronically while maintaining transportation-free tours. Hence, this motivates the scores from this ICT expert's research, where the maximum technology readiness is given to business and event tours. The usage trend during the COVID-19 pandemic presents support for positive support for both demand and the industry. Many ICTs are available for free, but paid programs are available that even can enable individuals to communicate in a virtual meeting room. Participants can appear as full-body avatars, replicating much of the body language that is often missed via regular video-conferencing software. Undoubtedly, demand for these services has also jumped dramatically since the start of COVID-19 [45].

Additionally, although cultural, urban, and entertainment types of tourism are less ready than business and event tourism, the results of this study indicate that adventure and rural tourism are far from ideal in terms of digitalization change readiness. Not only is there a high cost of technology for the satisfaction of the travel motivation of these groups, in proving the input data (with negative values) from expert panels, one article [48], for instance, analyzed the impact of real nature experiences against virtual nature experiences on well-being. Although the results show that interactive digital nature experiences may have comparable recovery effects to physical nature experiences, they offer only virtual reality where physical nature opportunities are limited, and there are many health benefits to aiming for a real walk in physical nature. The article argued that there also might be positive effects of light, physical activities (such as differences in seating and walking possibilities), and other moderating factors while traveling to real nature. Similarly, despite the existence of adventure virtual reality programs, a muscle-function analysis revealed that activation grades during such virtual reality programs were generally mild [49], which is not fully aligned with the travel motivation of adventure or rural tourists.

Last but not least, as the journal of the CiTUR Centre for Tourism Research, Development and Innovation recently expressed, contemporary and future tourism is expected to be dependent on two tendencies, development of technological innovations and sustainability [50]. This research is predicted to guide contributions to ICT-related innovations in tourism. It was stated [51] that ICT has the potential to lower travel costs, increase liquidity, and increase stability. It could also aid in the maintenance of social distancing in the pandemic, as ICT will link individuals again with no direct presence. As a result, this technology will deal with COVID-19-specific issues.

Now the public's confidence in this technology has grown, as has their ability to communicate and shift their attitudes toward technology. People have begun to disregard privacy concerns in order to reap greater technological benefits [51]. However, only those aspects of the tourism industry that recognize the benefits of ICTs and have effective management would be capable of improving their innovation and resistance [27]. Considering that theories on change management have highlighted the benefit of ensuring readiness for any change [52], as a roadmap for tourism strategists, this study is predicted to contribute to the concept of change management prior to formulating an action plan to encourage (or even discourage) a digitalization shift due to the COVID-19 crisis, post-COVID-19 trends, Fourth Industrial Revolution, environment and tourist attraction protection, or any other reason.

4. Conclusions

In response to the present tourism industry crisis, the purpose of this research was to comment on the readiness of tourist sector clusters for digital transformation. To do so, an expanded version of the TOPSIS technique was proposed to tackle MADM problems when working with altered and unbalanced inputs from expert panels. The proposed new approach converts the classic decision matrix to a multi-level, multi-panel, multi-criteria, and multi-alternative decision matrix. This expanded method was implemented to compare change readiness for transforming travel in tourism industry clusters for inbound and domestic Swedish tourism. Furthermore, due to a shortage of literature, a framework for measuring change readiness with a system perspective was adopted as an additional contribution to this work. Next, this research finding shows that business and event tourism can lead the transformation during and after the COVID-19 crisis. These two can better deal with the crisis because of their potential to serve the transition. However, adventure and rural tourism are the furthest away from being ready to adjust, and therefore suffer the most from the current crisis. Hence, it is to be expected that the findings of this study will assist authorities in assisting the industry with smarter decision- and strategy-making.

However, to fully understand the potential of the digitalization of travel, more studies are needed. Among the limitations of this study were the general questions asked about the readiness measures, as was commented on by one of the experts in response to the invitation to participate in this study. Hence, future works may use a qualitative approach to data collection through open-ended questions to bring more details to the analysis. When transitioning to digital tours, researchers may also suggest findings on the resilience management of tourism services (hotels, travel agents, etc.).

Funding: This research was funded by the Centre for Tourism (CFT) at Gothenburg University.

Institutional Review Board Statement: Not applicable.

Informed Consent Statement: Not applicable.

Data Availability Statement: Not applicable.

Conflicts of Interest: The author declares no conflict of interest.

Appendix A

Table A1 shows definition of the terminology used in this research.

Table A1. Operational definitions.

	Definition	Presented to Experts of:	
		Tourism	ICT
Digital tour/traveling	Any virtual (computer-generated) and/or online visits that reduce the need for travel and/or transportation	X	X
Demand response	Tourists' reaction to the digitalization shift for virtual and/or online travels	X	
Industry response	Industry (service providers of the industry) reaction to the digitalization shift for virtual and/or online travels	X	
Technology readiness	Availability of suitable technology infrastructure and knowledge to change to virtual and/or online travels at a reasonable price		X
Urban tourism	Includes visits to cities, towns, and the like	X	X
Cultural tourism	Travel to learn about other people, see architecture, art, history, etc.	X	X
Rural tourism	Undertakings in a non-urban territory, including coastal and nature tourism, stays in the countryside and rural retreats, national parks, etc.	X	X
Adventure tourism	It characteristically needs professional skills or physical exertion, and has some amount of risk.	X	X

Table A1. *Cont.*

	Definition	Presented to Experts of:	
		Tourism	**ICT**
Digital tour/traveling	Any virtual (computer-generated) and/or online visits that reduce the need for travel and/or transportation	X	X
Demand response	Tourists' reaction to the digitalization shift for virtual and/or online travels	X	
Industry response	Industry (service providers of the industry) reaction to the digitalization shift for virtual and/or online travels	X	
Technology readiness	Availability of suitable technology infrastructure and knowledge to change to virtual and/or online travels at a reasonable price		X
Urban tourism	Includes visits to cities, towns, and the like	X	X
Cultural tourism	Travel to learn about other people, see architecture, art, history, etc.	X	X
Rural tourism	Undertakings in a non-urban territory, including coastal and nature tourism, stays in the countryside and rural retreats, national parks, etc.	X	X
Adventure tourism	It characteristically needs professional skills or physical exertion, and has some amount of risk.	X	X
Event tourism	Attending any event or exhibition	X	X
Business tourism	Travel for business	X	X
Entertainment tourism	To enjoy entertainment activities, such as the circus, concerts, and clubbing	X	X
Travel motivation	Any specific reason, needs, or desires of tourists as the primary reason for traveling		X

Appendix B

The output from Decision Radar Ez-TOPSIS presented in Figure A1.

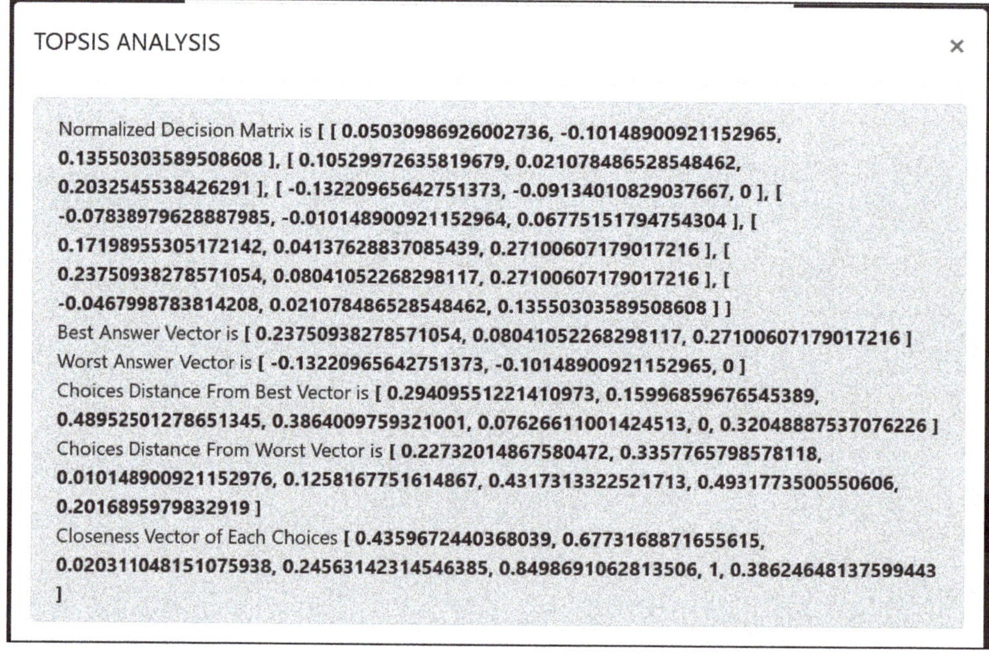

Figure A1. Software report.

References

1. Häfner, F.; Härting, R.-C.; Kaim, R. Potentials of digital approaches in a tourism industry with changing customer needs–a quantitative study. In Proceedings of the 15th Conference on Computer Science and Information Systems (FedCSIS), Sofia, Bulgaria, 6–9 September 2020.
2. FaladeObalade, T.A.; Dubey, S. Managing Tourism as a source of Revenue and Foreign direct investment inflow in a developing Country: The Jordanian Experience. *Int. J. Acad. Res. Econ. Manag. Sci.* **2014**, *3*, 16–42. [CrossRef]
3. Cura, F.; Singh, U.-S.; Talaat, K. Measuring the efficiency of tourism sector and the effect of tourism enablers on different types of tourism (Kurdistan). *Turizam* **2017**, *21*, 1–18. [CrossRef]
4. Shibata, N.; Shinoda, H.; Nanba, H.; Ishino, A.; Takezawa, T. Classification and Visualization of Travel Blog Entries Based on Types of Tourism. In *Information and Communication Technologies in Tourism 2020*; Springer: Cham, Switzerland, 2019; pp. 27–37.
5. Mikhailov, S.; Kashevnik, A. Tourist Behaviour Analysis Based on Digital Pattern of Life—An Approach and Case Study. *Futur. Internet* **2020**, *12*, 165. [CrossRef]
6. Sorooshian, S. Quarantine Decision due to Coronavirus Pandemic. *Electron. J. Gen. Med.* **2020**, *17*, em206. [CrossRef]
7. Peceny, U.S.; Urbančič, J.; Mokorel, S.; Kuralt, V.; Ilijaš, T. Tourism 4.0: Challenges in Marketing a Paradigm Shift. In *Consumer Behavior and Marketing*; IntechOpen: London, UK, 2020; pp. 1–19. [CrossRef]
8. Gretzel, U.; Fuchs, M.; Baggio, R.; Hoepken, W.; Law, R.; Neidhardt, J.; Pesonen, J.; Zanker, M.; Xiang, Z. E-Tourism beyond COVID-19: A call for transformative research. *Inf. Technol. Tour.* **2020**, *22*, 187–203. [CrossRef]
9. Higgins-Desbiolles, F. Socialising tourism for social and ecological justice after COVID-19. *Tour. Geogr.* **2020**, *22*, 610–623. [CrossRef]
10. Zhou, J.; Zogan, H.; Yang, S.; Jameel, S.; Xu, G.; Chen, F. Detecting Community Depression Dynamics Due to COVID-19 Pandemic in Australia. *IEEE Trans. Comput. Soc. Syst.* **2021**, 1–10. [CrossRef]
11. Turna, J.; Zhang, J.; Lamberti, N.; Patterson, B.; Simpson, W.; Francisco, A.P.; Bergmann, C.G.; Van Ameringen, M. Anxiety, depression and stress during the COVID-19 pandemic: Results from a cross-sectional survey. *J. Psychiatr. Res.* **2021**, *137*, 96–103. [CrossRef] [PubMed]
12. Galea, S.; Merchant, R.M.; Lurie, N. The Mental Health Consequences of COVID-19 and Physical Distancing: The Need for Prevention and Early Intervention. *JAMA Intern. Med.* **2020**, *180*, 817–818. [CrossRef] [PubMed]
13. Faisal, R.A.; Jobe, M.C.; Ahmed, O.; Sharker, T. Mental Health Status, Anxiety, and Depression Levels of Bangladeshi University Students During the COVID-19 Pandemic. *Int. J. Ment. Health Addict.* **2021**, 1–16. [CrossRef]
14. Holmes, E.A.; O'Connor, R.C.; Perry, V.H.; Tracey, I.; Wessely, S.; Arseneault, L.; Ballard, C.; Christensen, H.; Silver, R.C.; Everall, I.; et al. Multidisciplinary research priorities for the COVID-19 pandemic: A call for action for mental health science. *Lancet Psychiatry* **2020**, *7*, 547–560. [CrossRef]
15. Schuch, F.B.; Bulzing, R.A.; Meyer, J.; Vancampfort, D.; Firth, J.; Stubbs, B.; Grabovac, I.; Willeit, P.; Tavares, V.D.O.; Calegaro, V.C.; et al. Associations of moderate to vigorous physical activity and sedentary behavior with depressive and anxiety symptoms in self-isolating people during the COVID-19 pandemic: A cross-sectional survey in Brazil. *Psychiatry Res.* **2020**, *292*, 113339. [CrossRef] [PubMed]
16. Garfin, D.R. Technology as a coping tool during the coronavirus disease 2019 (COVID-19) pandemic: Implications and recommendations. *Stress Health* **2020**, *36*, 555–559. [CrossRef] [PubMed]
17. Laungani, R.; Chen, I.; Lui, C.; Gong, J.; Yoo, D.; Miyamoto, J. *The Impact of COVID-19 on Media Consumption across North Asia*; Nielsen.Com: New York, NY, USA, 2020.
18. Brouder, P.; Teoh, S.; Salazar, N.B.; Mostafanezhad, M.; Pung, J.M.; Lapointe, D.; Desbiolles, F.H.; Haywood, M.; Hall, C.M.; Clausen, H.B. Reflections and discussions: Tourism matters in the new normal post COVID-19. *Tour. Geogr.* **2020**, *22*, 735–746. [CrossRef]
19. Tarman, B. Editorial: Reflecting in the shade of pandemic. *Res. Soc. Sci. Technol.* **2020**, *5*. [CrossRef]
20. Chamarro, A. Psychosocial impact of COVID-19: Some evidence, many doubts to be clarified. *Aloma Rev. Psicol. Ciències l'Educació i l'Esport* **2020**, *38*, 9–10. [CrossRef]
21. Kolev, Y. Virtual Tours-Tourism during COVID-19 Crisis. 2020. Available online: https://www.bnr.bg/en/post/101247369/virtual-tours-tourism-during-covid-19-crisis (accessed on 29 March 2020).
22. Navigating the New Normal: The Future of Travel in a Post-COVID-19 World. *Travel Daily News*, 25 May 2020.
23. Monica, O.; Elena, T. Durable development of tourism and its impact on the environment. *Anale. Ser. Stiinte Econ. Timis.* **2012**, *18*, 156–160.
24. Haley, A.; Snaith, T.; Miller, G. The social impacts of tourism a case study of Bath, UK. *Ann. Tour. Res.* **2005**, *32*, 647–668. [CrossRef]
25. Dolnicar, S. Understanding barriers to leisure travel: Tourist fears as a marketing basis. *J. Vacat. Mark.* **2005**, *11*, 197–208. [CrossRef]
26. Buhalis, D.; Deimezi, O. E-Tourism Developments in Greece: Information Communication Technologies Adoption for the Strategic Management of the Greek Tourism Industry. *Tour. Hosp. Res.* **2004**, *5*, 103–130. [CrossRef]
27. Buhalis, D.; O'Connor, P. Information Communication Technology Revolutionizing Tourism. *Tour. Recreat. Res.* **2005**, *30*, 7–16. [CrossRef]
28. Ozturk, H.M. Technological Developments: Industry 4.0 and Its Effect on the Tourism Sector. In *Handbook of Research on Smart Technology Applications in the Tourism Industry*; IGI Global: Hershey, PA, 2020; pp. 205–228.

29. Pencarelli, T. The digital revolution in the travel and tourism industry. *Inf. Technol. Tour.* **2020**, *22*, 455–476. [CrossRef]
30. Hayes, J. *The Theory and Practice of Change Management*; Palgrave: London, UK, 2018.
31. Napierała, T.; Bahar, M.; Leśniewska-Napierała, K.; Topsakal, Y. Technology towards hotel competitiveness: Case of Antalya, Turkey. *Eur. J. Tour. Hosp. Recreat.* **2020**, *10*, 262–273. [CrossRef]
32. Heizer, J.; Render, B.; Munson, C. *Operations Management: Sustainability and Supply Chain Management*; Pearson: Hoboken, NJ, USA, 2017.
33. Sorooshian, S.; Azizan, N.A.; Ismail, M.Y. Influence of readiness measures on planning tourism digital shift. *Acad. Strateg. Manag. J.* **2021**, *20*, 78.
34. Wang, S. Museums' Responses to Tourism Sessonality: A Case Study in Umeå, Sweden. Master's Thesis, Umeå University, Umeå, Sweden, 2019.
35. Városiné Demeter, K.; Losonci, D.; Szász, L.; Rácz, B.G. Assessing Industry 4.0 readiness: a multi-country industry level analysis. In Proceedings of the 25th Annual EurOMA Conference, Budapest, Hungary, 24–26 June 2018.
36. Andersson, T.D.; Getz, D. Festival Ownership. Differences between Public, Nonprofit and Private Festivals in Sweden. *Scand. J. Hosp. Tour.* **2009**, *9*, 249–265. [CrossRef]
37. Abdulhadi, A.Q. Review of Hybrid TOPSIS with other Methods. *Int. J. Acad. Res. Bus. Soc. Sci.* **2019**, *9*, 52–62. [CrossRef]
38. Yue, Z. Approach to group decision making based on determining the weights of experts by using projection method. *Appl. Math. Model.* **2012**, *36*, 2900–2910. [CrossRef]
39. Shih, H.-S.; Shyur, H.-J.; Lee, E.S. An extension of TOPSIS for group decision making. *Math. Comput. Model.* **2007**, *45*, 801–813. [CrossRef]
40. Widianta, M.M.D.; Rizaldi, T.; Setyohadi, D.P.S.; Riskiawan, H.Y. Comparison of Multi-Criteria Decision Support Methods (AHP, TOPSIS, SAW & PROMENTHEE) for Employee Placement. *J. Phys. Conf. Ser.* **2018**, *953*, 12116. [CrossRef]
41. Handels, R.L.; Wolfs, C.A.G.; Aalten, P.; Bossuyt, P.M.M.; Joore, M.A.; Leentjens, A.F.G.; Severens, J.H.; Verhey, F.R.J. Optimizing the use of expert panel reference diagnoses in diagnostic studies of multidimensional syndromes. *BMC Neurol.* **2014**, *14*, 1–9. [CrossRef] [PubMed]
42. Keshtkar, M.M. Performance analysis of a counter flow wet cooling tower and selection of optimum operative condition by MCDM-TOPSIS method. *Appl. Therm. Eng.* **2017**, *114*, 776–784. [CrossRef]
43. Sorooshian, S.; Parsia, Y. Modified Weighted Sum Method for Decisions with Altered Sources of Information. *Math. Stat.* **2019**, *7*, 57–60. [CrossRef]
44. Sorooshian, S. Group Decision Making With Unbalanced-Expertise. *J. Phys. Conf. Ser.* **2018**, *1028*, 012003. [CrossRef]
45. Gray, L.; Wong-Wylie, G.; Rempel, G.; Cook, K. Expanding Qualitative Research Interviewing Strategies: Zoom Video Communications. *Qual. Rep.* **2020**, *25*, 1292–1301. [CrossRef]
46. Arshad, M. COVID-19: It's time to be thankful to our ICT professionals. *Inf. Technol. Electr. Eng.* **2020**, *9*, 23–31.
47. Wiederhold, B.K. Connecting Through Technology During the Coronavirus Disease 2019 Pandemic: Avoiding "Zoom Fatigue". *Cyberpsychol. Behav. Soc. Netw.* **2020**, *23*, 437–438. [CrossRef] [PubMed]
48. Reese, G.; Stahlberg, J.; Menzel, C. Digital shinrin-yoku: Do nature experiences in virtual reality reduce stress and increase well-being as strongly as similar experiences in a physical forest? *Eff. Virtual Environ. Well-Being* **2021**. [CrossRef]
49. De Vries, A.W.; Willaert, J.; Jonkers, I.; van Dieën, J.H.; Verschueren, S.M.P. Virtual reality balance games provide little muscular challenge to prevent muscle weakness in healthy older adults. *Games Health J.* **2020**, *9*, 227–236. [CrossRef]
50. Napierała, T.; Birdir, K. Competition in Hotel Industry: Theory, Evidence and Business Practice. *Eur. J. Tour. Hosp. Recreat.* **2020**, *10*, 200–202. [CrossRef]
51. Sharma, G.D.; Thomas, A.; Paul, J. Reviving tourism industry post-COVID-19: A resilience-based framework. *Tour. Manag. Perspect.* **2021**, *37*, 100786. [CrossRef]
52. Weiner, B.J. A theory of organizational readiness for change. *Implement. Sci.* **2009**, *64*, 1–9. [CrossRef] [PubMed]

Article

Modelling the Enablers for Branded Content as a Strategic Marketing Tool in the COVID-19 Era

Shilpa Sindhu [1] and Rahul S Mor [2],*

[1] School of Management, The NorthCap University, Gurugram 122017, India; shilpasindhu@ncuindia.edu
[2] Department of Food Engineering, National Institute of Food Technology Entrepreneurship and Management (NIFTEM), Kundli, Sonepat 131028, India
* Correspondence: dr.rahulmor@gmail.com

Abstract: This study aims towards identifying and modelling the significant factors which act as enablers for the branded content to be used strategically by marketers as a marketing tool in the COVID-19 era. A qualitative approach was adopted for this study, and significant factors associated with branded content were identified from the literature review and primary survey. The factors were then verified by the experts in the area of branding and digital marketing. Total interpretive structural modelling (TISM) and Decision-making Trial and Evaluation Laboratory (DEMATEL) techniques were used to model the factors as per their contextual relationships. As per the model outcomes from TISM and DEMATEL approaches, branded content is an efficient marketing tool that promises value delivery to stakeholders. This, in turn, depends on the authenticity and transparency in content development and distribution. The most significant driving enablers for the system suggest efficient measurement and evaluation strategies and the customer as co-creator for the branded content.

Keywords: branded content; marketing; total interpretive structural modelling; decision-making Trial and Evaluation Laboratory

Citation: Sindhu, S.; Mor, R.S. Modelling the Enablers for Branded Content as a Strategic Marketing Tool in the COVID-19 Era. *Systems* **2021**, *9*, 64. https://doi.org/10.3390/systems9030064

Academic Editors: Oz Sahin and Russell Richards

Received: 12 July 2021
Accepted: 20 August 2021
Published: 26 August 2021

Publisher's Note: MDPI stays neutral with regard to jurisdictional claims in published maps and institutional affiliations.

1. Introduction

Marketing as a concept and practice has embraced various definitions and approaches. With the change in time and versatile consumer behaviour, marketing tools adopted by marketers demand responsiveness. The visible indication for this is seen in the recent outbreak of pandemic COVID-19, which has almost shaken the full dynamics of marketing. There is a sudden upsurge in the quantum of online buyers in the year 2020. A few years back, online retailing contributed only around 3%of total retail in India, which is now expected to be around 8% in 2021 [1,2]. Marketers have observed record hit for search on the internet for their products or services. Digital platforms have become necessary for almost every marketer due to the upsurge of digital content and various entrants. The content used for product promotion may be created by the marketer itself or developed through any other company or user-generated. The content thus generated is called branded content, which may be defined as "any output fully/partly funded or at least endorsed by the legal owner of the brand which promotes the owner's brand values, and makes audiences choose to engage with the brand based on a pull logic due to its entertainment, information or education value" [3]. The term branded content is not new and is being used across all major continents globally through close association with Branded Content Marketing Association (BCMA) for over a decade [4]. Marketers try to reach as close to the consumer as possible to promote their products by providing brand information through user-generated or brand-generated content [5].

Branded content may be a video, an article, an audiovisual, a blog, a magazine, or an event, etc. [3]. For this study, branded content created by the marketers themselves or through outsourcing in digital form is considered for discussion. It is different from traditional advertising, where the promotional content comes as a push factor, intermitting with the customer's program of choice. Instead, digital branded content is used as a pull factor, where a separate video or digital content is there, with the aim of least obtrusion in the customer's running show. The customer decides to watch or read branded content out of choice rather than out of compulsion. But this does not mean that branded content intends to replace traditional advertisements; instead, they both can be used strategically where traditional advertisements focus on sales and branded content focus on brand image and brand extension [4]. In other words, digital branded content is any digital form of media that is "intentional, brand-authored media used to establish or extend brand identity or affinity" [6]. It is also established from the concept of permission marketing that consumers don't like to be disturbed by promotional content without their choice [7]. The customer, therefore, may feel more associated with the brands of their choice through branded content and content delivery. The experience of branded content acts as a driver for customer engagement with the brand and gives a feeling of virtual association [8]. Creating and managing the correct branded content which can promote the brands remains a challenge for marketers always [9].

The pandemic period has allowed marketers to exploit the opportunity of bringing the product closer to consumers through branded content. While buying from home, the consumer looks for product reviews, feedbacks, and product details deeply. A branded content revolves broadly around five areas viz. choice, deliverable vs. discipline, engagement, mode of delivery, and value [3]. Scanty research is available on understanding all the aspects of branded content as a marketing strategy. As per a few experts, the effects of a pandemic may go as long as up to the year 2022. After that, whether marketing will begin as before the COVID-19 period or whether it will be a new normal is a question to ponder upon, and the trend shows more chances for getting the new normal trend to be set up. Various researchers have emphasized that being innovative digitally can help companies survive this pandemic. During this pandemic, more companies have started using WhatsApp, Google Meet, Zoom, etc., reflecting the path ahead [10]. In such a scenario, marketers need to focus deeply and strategically on branded content as a competitive strategy. This is where the need for this study is realized.

This study is built upon the following research questions:

R1. Which are the major enabling factors for the branded content to be used strategically by marketers as a marketing tool.

The study aims to highlight the major factors that can act as enablers for branded content success.

R2. What should be a guiding framework for long-term sustainability for branded content as a strategic marketing tool, based on the interaction among identified factors?

Not all the factors need equal focus and efforts from marketers; therefore, an effort has been made to model the factors in a meaningful and contextual hierarchical relationship. The qualitative approach for data collection and analysis was approached, where factors were identified through extensive literature review and primary data and were validated through expert opinion. The factors were then modelled through Total Interpretive Structural Modelling (TISM) and Decision-making Trial and Evaluation Laboratory (DEMATEL) Techniques. This study emphasizes strategically approaching the branded content to be used as a marketing tool in the COVID-19 scenario and lays the foundation for marketers and policymakers.

This study further has sections, where Section 2 presents the literature review, and Section 3 discusses the methodology, including TISM and DEMATEL techniques. Section 4 is the results and analysis, and Section 5 presents the discussion and practical implications. Section 6 gives the conclusion, and Section 7 is the limitations and future research directions for this study.

2. Literature Review

The concept of online presence or digital marketing is not new for marketers or consumers. Digital marketing is there for years, and marketers adopt everyday innovative approaches to make close bonds with consumers [4] and studying the consumer attitude and behaviour in the technology-driven sharing economy [11–13]. Consumers are targeted through social media channels, display advertisements, and search engines [14–16]. Online social platforms are most interactive and communicative to reach consumers [17]. Various retail marketers are using social media platforms to promote their brands, often taking the help of brand communities on social media platforms [18–20]. Research suggest that, if designed and directed by marketers for brand promotion, social media platforms can develop a sense of belongingness and interactivity among consumers [21]. Branded content is relatively still a new topic [4]. Branded content is developed to provide focused information entertainingly to a specific target set of consumers. Influencer-marketed branded content is prominent, wherein the marketer chooses a personality to create and market its branded content [19,22]. The content needs to be vivid and deeply connected with the products [20]. Effective content formats also impact influencer marketing [23].

The branded content development and distribution have become a buzz feature for marketers nowadays, which is discussed widely with content marketing [3]. The terms branded content, and branded content marketing should be used carefully as both carry a separate meaning and require separate attention from marketers. In one way, branded content marketing may be considered an enabler for making the content reach its target customers. The key to using branded content efficiently lies in the adequately defined content and carefully selected medium to make it available to consumers. The Branded Content Marketing Association (BCMA) suggested a few rules for branded content success as a marketing tool. The suggestions include that interesting and original branded content, promotion of content, combined campaigns that can enhance the positive image of a brand, and synergistically with traditional advertising [6]. Authors tried to sensitize the readers about the benefits of relating branded content with crowd culture through social media [24]. Branded content integration may also be done by either placing the product or brand in the movie or TV program, or it may be used to portray the real use by using any celebrity [25]. Webisode communication is also a branded content strategy wherein movies or series are broadcasted on the internet related to a brand [26].

The significant factors as enablers for branded content to be used as an efficient marketing tool may be discussed as follows:

2.1. Factor 1: Branded Content Distribution and Promotion Strategy

Branded content must be delivered and distributed to the consumers most effectively once created. The distribution of the promotion aspect of branded content is discussed under the umbrella of content marketing. An efficient content marketing influences consumers' purchase intentions and loyalty to brands [27]. Consumers tend to have different brand perceptions based on how the content is promoted and presented to consumers [4]. The time of content delivery impacts its popularity [28]. There are various channels available for branded content delivery, so choosing the best content delivery method is a crucial for marketers these days [6,29].

2.2. Factor 2: Quality of Content

The branded content needs to be creative, informative, and engaging. Marketers use different forms of content to engage the consumers and pass the brand message in a more focussed manner [6,27,30]. Consumers look for different types of content for different product types, like complete information and advice for health products and automobiles, while inspirational content for fashion products [19]. The type and quality of content decide the engagement power of consumers with the brand [31]. The richness of the content and proper usage of images impact the popularity of branded content amongst the consumers [28].

2.3. Factor 3: Authenticity

Consumers consult the branded content to get adequate information about the brand. Therefore, the credibility of branded content is an important decisive factor for consumer choice [32]. Content authenticity affects the brand image directly by impacting consumer trust. Consumers often relate the authenticity of any engagement initiative with brand perception; if the initiative matches with perception, it is considered authentic [33]. The authenticity of branded content depends to a great extent on the source and mode of content development.

2.4. Factor 4: Transparency

The amount of disclosure in the branded content relates to transparency and ethics [34]. Consumers expect brand marketers to be transparent in their content descriptions and delivery methods, but marketers need to consider various business decisions. Therefore, the level of transparency in branded content is tricky, and it needs to be managed with due diligence. Lack of transparency also may disorient the consumers and may lead to a lack of trust in the brands. As per the study conducted [35], companies may introduce disclaimers with new formats of creating branded content to protect the credibility and transparency of brands. The advertising of branded content also needs due consideration and compliance with the legislative mechanism for bringing in required transparency [36].

2.5. Factor 5: Value Delivery to Stakeholders

Branded content success as a marketing tool depends on how it is embraced by consumers [3]. Also, besides consumers, branded content should provide value to its other stakeholders that prominently involve the marketer company and the company if hired to develop branded content on behalf of the marketer. If the branded content can provide value to the stakeholders, it can only sustain the competition and prove a competitive tool for the marketer. There needs to be an affinity between the marketer and the media providers for delivering branded content [37].

2.6. Factor 6: Measurement and Evaluation of Branded Content

As per [6], the branded content needs to be evaluated for quality, and its impact needs to be measured for its success. BCMA suggests a content evaluation system known as Branded Content Evaluation System (BSES), which focuses on overall content performance, component-wise impact, and anything the marketer needs to do differently from the competitors. Many companies face difficulties related to content production and curation and ROI measurements [38].

2.7. Factor 7: Customer as Co-Creator

The branded content gets enriched when the consumer becomes one of the participants in creating it. Marketers are exploring online communities of consumers as a tool for brand co-creation [39]. Marketers invite consumers to participate in the co-creation of brand content through various means [40,41]. Studies claim that consumers usually believe the reviews or content provided by other consumers more than those professionally created [4]. Also, if inviting suggestions from customers, the marketers should respond or attend to those suggestions or concerns to have content and product improvements [6]. Table 1 presents a summary of the literature.

Table 1. Summary of Literature Support for the Identified Factors.

Sr. No	Factors	Literature Support
1	Content distribution and promotion strategy	[4,6,25,27–29]
2	Quality of content	[6,19,20,22,27,28,30,31]
3	Authenticity	[32]
4	Transparency	[34]
5	Value delivery to stakeholders	[3,37]
6	Measurement and evolution of branded content	[6,38]
7	Customer as co-creator	[4,6,39,40]

3. Methodology

This study aims to identify the significant factors that may act as enablers for branded content and be strategically used as a marketing tool by marketers in the COVID-19 era. For this purpose, a qualitative approach was adopted. As a first step, literature was screened to identify the significant factors associated with branded content. The identified factors (14 factors) were then randomly circulated online in the last week of June 2020 to a few respondents for responses on factor appropriateness. A brief questionnaire was prepared for this purpose and posted on social media platforms like Watsapp and LinkedIn. The responses were received from 83 respondents. A very brief summation of the same is mentioned in Table A2 (Appendix B). The final list of factors was prepared based on responses received and the literature survey (as mentioned in section two above). The factors were then verified by the experts in the area of branding and digital marketing. Also, the experts were asked to identify the contextual relationship between the variables as per the requirements of the TISM and DEMATEL approach. A total of five experts were approached for this purpose, including three academicians in leading management institutes in Delhi-NCR, and two were from the industry. The domain experts were chosen based on their Linkedin profile and experience in the subject. For response collection purposes, all the five experts were called on a virtual meeting two times during 2nd July to 13th July 2020, and the researcher recorded their observations and comments. Based on the suggestions from the experts, TISM and DEMATEL approaches were applied. Both these approaches have demonstrated their strength for modelling variables in different research domains. TISM is an extension of the Interpretive Structural Modeling (ISM) approach. It is preferred over ISM, as it overcomes a few of the drawbacks of the ISM approach such as in ISM no logic is provided for the identified relationships between the variables. In contrast, in TISM, the interpretive logic knowledge base matrix is prepared to provide logic for each linkage. TISM approach identifies the structural relationship between the variables [42–45], and DEMATEL further provides the strength of those relationships, along with providing the cause and effect relationship among the variables [46–48]. Further, in this section, the methodology for TISM and DEMATEL approaches are discussed.

3.1. Total Interpretive Structural Modelling (TISM)

Total interpretive structural modelling (TISM) is a qualitative approach for identifying the contextual relationships between the factors under study [49,50]. TISM highlights the driving power or dependence of one factor over others and thereby identifies the significant linkages. The steps involved in TISM [51–53] are discussed as below:

Step 1: Identification of relevant factors from the literature review and validation from experts.

Step 2: Developing the interpretive logic-knowledge base matrix for the contextual relationship ("lead to" type) amongst the factors, as per inputs from the experts, and marking the entries as YES or NO. Wherever one particular factor leads to another factor, entry is made as 'YES' in the matrix, and wherever the 'lead to' relation is missing, entry

113

is made as 'NO' in the matrix. Also, the experts are asked to provide a logical reason for the proposed relationship between the factors, and the reason is mentioned in the column against the 'YES' entry.

Step 3: Transformation of the interpretive logic-knowledge base matrix into a binary matrix (reachability matrix) by making (i,j) entry of YES as (i,j) entry of 1 in the reachability matrix, and (i,j) entry of NO as (i,j) entry of 0 in the reachability matrix. The reachability matrix is then scrutinized for transitive links as per the following formula:

'If factor1 leads to factor 2, and factor 2 leads to factor 3, then factor 1 should also lead to factor 3'and wherever it is found, transitivities are included in the form of 1* in the reachability matrix.

Step 4: The factors differ in their magnitude and direction to influence other factors in the system. In the TISM approach, factor level partitioning is done to allot level to each factor as per its magnitude and direction of influence. The step of allotting level to any factor is called iteration. For each iteration, the reachability set (consists of all the factors that this factor leads to, including self) and the antecedent set (consists of all the factors that lead to this factor, including self) are identified intersections are recorded under the intersection set. Levels are allotted to any factor whenever its reachability and intersections set becomes equal, and then that specific factor is removed from further iterations. Iterations in this way continue till levels are allotted to each factor.

Step 5: Carrying out MICMAC analysis for grouping the factors into four clusters viz. 'autonomous', 'dependent', 'linkage' and 'drivers' based on their driving power and dependence.

Step 6: Preparing the diagraph/TISM model, based on the levels achieved by each faculty, to represent the direction of influence of one factor on another graphically.

3.2. DEMATEL

The DEMATEL technique is used for developing and interpreting the causal or effect relationship between the identified factors [54]. The DEMATEL approach consists of the following steps [55–59].

Step 1: Developing the direct relation matrix (D): To develop the direct relation matrix (D), the pair-wise relationships amongst the factors are established first. For this purpose, expert opinion is sought for evaluating all the pairs of factors on a scale of 0–4, where value 0 denotes 'no influence' and value 4 denotes 'extreme strong influence' of one factor on another in the pair. Accordingly, a non-negative matrix (n × n) is achieved for n factors for each expert. After that, the responses of all the experts are averaged and accordingly, the direct relation matrix (D) is obtained by following the below-mentioned formula:

$$D = \frac{1}{n} \sum_{k=1}^{n} D_{ij}^{k}$$

Step 2: Developing the normalized direct relation matrix (N): The normalized direct relation matrix (N) is obtained by normalizing the direct relation matrix (D), by using the formula:

$$N = D/k$$
$$K = \max_{i,j} \left(\max_i \sum_{j=1}^{n} a_{i,j} \max_j \sum_{i=1}^{n} a_{i,j} \right), \; i, j = 1, 2, 3$$

Step 3: Developing the total relation matrix (R)
Total relation matrix (R) is obtained from the normalized matrix by using the formula:

$$R = N(I - N)^{-1}$$

where I represent the identity matrix. The total relation matrix (R) depicts the type of relationship (influence); one factor has over other factors.

Step 4: Developing the causal diagram: The causal diagram in DEMATEL is obtained by plotting the values of (D+R) and (D−R), where 'D' denotes the sum of rows, and 'R' denotes the sum of columns, respectively, for each factor. The value of the sum of rows ('D') shows the sum of the influence of one particular factor on other factors, and the value of the sum of columns ('R') shows the sum of the influence of other factors on that respective factor. Similarly, values of (D+R) reflects the strength of the relationship of the particular factor with the system. Similarly, values of (D−R) reflect the nature of relationships amongst all the factors. The positive value of (D−R) of the factor shows that the respective factor belongs to the cause group. The negative value of (D−R) of factor indicates that the respective factor belongs to the effect group. The (D+R) and (D−R) values are further plotted on the x and *y*-axis to obtain the causal diagram.

4. Results and Analysis

4.1. TISM Modelling

Step-wise results from TISM modelling are discussed as follows.

Step 1: Identification and listing of the relevant factors: A total of seven factors were identified from the literature review and primary survey, which may act as enablers for branded content to be used as a marketing strategy. All seven factors are described in Section 2 above. Also, the factors were verified by the experts, as mentioned in Section 3 beginning.

Step 2: Defining Contextual Relationship and developing an Interpretive logic-knowledge base. The Interpretive logic-knowledge base matrix was prepared as per the methodology in step 2 in Section 3 above and is placed as Table A1 (Appendix A).

Step 3: Development of a reachability matrix from the Interpretive logic-knowledge base and then scrutinize the matrix for transitivity. The interpretive logic-knowledge base was transformed into a binary matrix following the process described in step 3 in Section 4 above. The reachability matrix obtained is placed in Table 2. Further, as per the rule for transitivity discussed in step 3 in Section 4 above, the final reachability matrix is prepared and placed in Table 3. Also, the transitivities such obtained were included in the interpretive logic-knowledge base (Table A1), by replacing the entry of NO with the entry of YES, for that respective transitive entry and also the word 'transitive' was written in the respective column of that entry. Further, the driving power (calculated by adding up the number of 1s in the row) and dependence (calculated by adding up the number of 1s in the column)for each factor were calculated and recorded in the final reachability matrix.

Step 4: Carrying out level partitioning of the reachability matrix: As per the process of level partitioning detailed in step 4 of Section 3 above, In this study, a total of four iterations were required to allot levels to each factor. The consolidated level partition table is placed as Table 4.

Step 5: MICMAC Analysis: The purpose of MICMAC Analysis is to divide the identified factors into four different clusters as per the driving power and dependence of the factors. The four clusters thus identify and group the factors as autonomous, dependent, linkage, and independent factors [44,60–63]. The grouping of factors in this study is presented in Figure 1 and discussed as below:

Table 2. Initial Reachability Matrix.

(i,j)	1	2	3	4	5	6	7
1	1	0	1	1	1	0	0
2	0	1	1	0	1	0	0
3	0	0	1	1	1	0	0
4	0	0	1	1	1	0	0
5	0	0	0	0	1	0	0
6	1	1	1	1	0	1	0
7	1	1	0	1	1	0	1

Table 3. Final Reachability Matrix(Transitivity).

(i,j)	1	2	3	4	5	6	7	Driving Power
1	1	0	1	1	1	0	0	4
2	0	1	1	1 *	1	0	0	4
3	0	0	1	1	1	0	0	3
4	0	0	1	1	1	0	0	3
5	0	0	0	0	1	0	0	1
6	1	1	1	1	1 *	1	0	6
7	1	1	1 *	1	1	0	1	6
Dependance	3	3	6	6	7	1	1	

* Transitivity.

Table 4. Consolidated Level of Factors.

Factors	Reachability Set	Antecedent Set	Intersection Set	Level
1	1,3,4,5	1,6,7	1	3
2	2,3,4,5	2,6,7	2	3
3	3,4,5	1,2,3,4,6,7	3,4	2
4	3,4,5	1,2,3,4,6,7	3,4	2
5	1	1,2,3,4,5,6,7	1	1
6	1,2,3,4,5,6	6	6	4
7	1,2,3,4,5,7	7	7	4

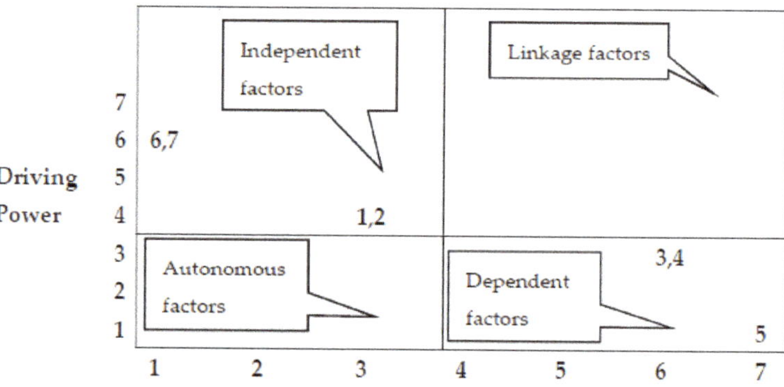

Figure 1. MICMAC Analysis.

Cluster I: This cluster groups together Autonomous Factors in the system. Such factors do not significantly relate to other factors and have weak driving power and weak dependence. In this study, no factors emerged into this group, reflecting that all the factors show some other types of relationships.

Cluster II: This cluster groups together Dependent Factors in the system. Such factors have weak driving power and high dependence on other factors. In this study, value delivery to stakeholders (5), transparency (4), and authenticity (3) were grouped into this cluster. These factors are strategic for the system but need the support of other factors to be achieved successfully.

Cluster III: This cluster groups togetherLinkage Factors in the system. Such factors have driving power and dependence both as high. They are the most unstable ones, and any change on other factors can easily reflect on these factors and other factors. In this study, no factor emerged as a linkage factor, which might be because all the identified factors have either significant driving power or dependence, but not both.

Cluster IV: This cluster groups together Independent Factors in the system. Such factors have high driving power and low dependence on other factors. In this study, measurement and evaluation strategies (6), the customer as co-creator (7), quality of the content (2), and distribution and promotion strategy (1) got categorized into this cluster.

Step 6: TISM Model/Diagraph

All the factors are represented graphically in the sequence as per their driving powers and dependence, and the model thus obtained is known as the TISM model or Diagraph. In this study, the seven factors were placed as per their level partitions, where the factor with level one was placed at the top, followed by next-level factors. Factors are connected through arrows, which always point upward in vertical interrelationships, and arrows point to both sides in case of horizontal or same level factors. The dotted lines in the model reflect the indirect 'lead to' relation between the factors. The TISM model so generated is placed as Figure 2.

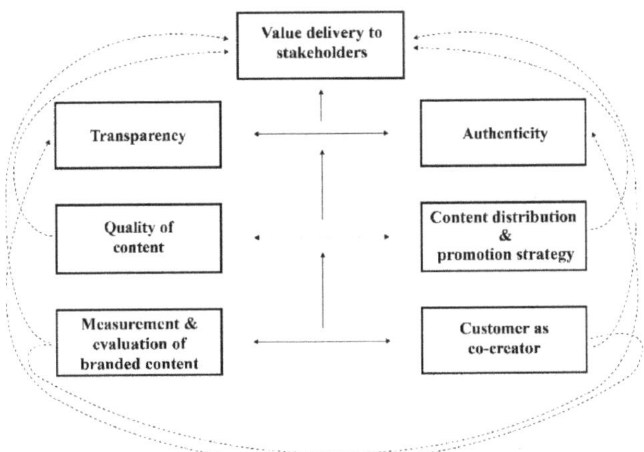

Figure 2. TISM Model.

As per the TISM model, the factors viz. measurement and evaluation strategies (6); and the customer as co-creator (7) emerged as the most significant driving forces for other factors. This signifies that for branded content to be used strategically for brand promotion, it is essential to have well-designed and practice measurement and evaluation strategies. These measurement and evaluation strategies help keep the quality of content higher and its delivery platforms efficient. The marketer needs to draft methods to measure the impact of branded content on the marketing and promotion aspects of their products. Similarly, the content development and delivery are to be evaluated frequently on pre-decided parameters with much precision. Equally, the strong enabler is the customer as co-creator (7), which also has high driving power. This is realized that if branded content focuses on and usage of user-generated content, its impact becomes manifold. Both these driving forces are significant to maintain transparency and credibility in the system for all the stakeholders. As per the model, next in the hierarchy in driving power are the quality of the content (2); and distribution and promotion strategy (1). With ideal measurement and evaluation strategies and involving the customer as content co-creator, it helps develop excellent quality content and content distribution in the most user-acceptable manner.

Consumers seek complete and accurate information about the product before buying, and that is what they expect the branded content should provide them. Also, not all ways of making the content available to consumers are effective. Marketers need to identify the delivery method, which suits their product type and the profiling of consumers in the best suitable manner. These four driving forces discussed above lead to branded content's transparency (4), and authenticity (3) for the customers.

Consequently, all these factors lead to value delivery to stakeholders (5). As emerged from the model, value delivery to stakeholders eventually decides the fate of the system's success. Stakeholders involve the customers, content developers, platform providers, and marketers (if different from content creators and distributors). Unless the stakeholders get something worth higher sales, better product reviews, acceptance, and increased profits, only branded content should be sustained as a promotion strategy.

4.2. DEMATEL Model

The methodology explained in Section 3.2 above was adopted, and consequently, the step-wise results obtained from applying the DEMATEL technique on the factors are discussed below:

Step 1: Direct relation matrix (D)

The direct relation matrix (D) was developed by identifying the pair-wise relationship between the identified factors, as per the method and formula explained in step 1 of Section 3.2. The direct relation matrix is placed in Table 5.

Table 5. Direct Relation Matrix (D).

(i,j)	1	2	3	4	5	6	7
1	0	0	4	4	4	0	0
2	0	0	3	4	4	1	0
3	0	1	0	3	4	1	1
4	2	3	4	0	4	1	2
5	1	3	4	4	0	2	2
6	4	4	4	4	4	0	4
7	3	4	4	4	4	3	0

0 = No influence; 1 = Low influence; 2 = Medium influence; 3 = High influence; 4 = Very High influence.

Step 2: Normalised direct relation matrix (N)

The normalized direct relation matrix (N) was obtained by normalizing the direct relation matrix (D) using the formula mentioned in step 2 in Section 3.2 above. Accordingly, the matrix obtained is mentioned in Table 6.

Table 6. Normalized direct relation matrix (N).

(i,j)	1	2	3	4	5	6	7
1	0.00000	0.00000	0.16667	0.16667	0.16667	0.00000	0.00000
2	0.00000	0.00000	0.12500	0.16667	0.16667	0.04167	0.00000
3	0.00000	0.04167	0.00000	0.12500	0.16667	0.04167	0.04167
4	0.08333	0.12500	0.16667	0.00000	0.16667	0.04167	0.08333
5	0.04167	0.12500	0.16667	0.16667	0.00000	0.08333	0.08333
6	0.16667	0.16667	0.16667	0.16667	0.16667	0.00000	0.16667
7	0.12500	0.16667	0.16667	0.16667	0.16667	0.12500	0.00000

Step 3: Total relation matrix (T)

The total relation matrix (T) obtained as per the formula mentioned in step 3 of Section 3.2 above is placed in Table 7.

Table 7. Total relation matrix (T).

(i,j)	1	2	3	4	5	6	7
1	0.06131	0.11857	0.32363	0.31631	0.32787	0.06875	0.07862
2	0.07056	0.12810	0.29613	0.32584	0.33642	0.11172	0.08615
3	0.06824	0.15930	0.16650	0.27339	0.31505	0.10733	0.11553
4	0.16357	0.27125	0.38438	0.23748	0.39407	0.13350	0.17423
5	0.13603	0.28087	0.39091	0.38698	0.25808	0.17170	0.18199
6	0.29029	0.38800	0.50638	0.50215	0.52024	0.13855	0.29606
7	0.24202	0.36991	0.47674	0.47292	0.48995	0.23829	0.13982

Step 4: Developing the causal diagram based on values of (D+R) and (D−R):

From the total relation matrix, the values of (D+R), i.e., the sum of influences given to factors, and (D−R), i.e., the sum of influences received by factors, were calculated, as shown in Table 8.

Table 8. The sum of influences (given to and received by) the factors.

Factors	D	R	(D+R)	(D−R)
1	1.2950615	1.032015	2.327076437	0.26304664
2	1.3549088	1.715992	3.070900834	−0.361083279
3	1.205329	2.544663	3.749992006	−1.339334033
4	1.7584737	2.515068	4.273541299	−0.756593956
5	1.8065666	2.641663	4.448229307	−0.835096164
6	2.6416627	0.969843	3.61150587	1.671819602
7	2.4296369	1.072396	3.502032547	1.35724119

The factors were ranked based on their (D+R) values, reflecting the relative importance of the factor in the system and the degree of the relation of one factor with other factors. The same is highlighted in Table 9.

Table 9. The relationship strength rankings.

Ranks	Factor	(D+R)
1	5	4.448229
2	4	4.273541
3	3	3.749992
4	6	3.611506
5	7	3.502033
6	2	3.070901
7	1	2.327076

Similarly, the factors were also ranked based on their values of (D−R), reflecting the kind of relation between the variables and summarised in Table 10.

Table 10. The relation type and relative rankings.

Ranks	Cause-Group Factor (+ve Value of D−R)	(D−R)
1	6	1.671819602
2	7	1.35724119
3	1	0.26304664
Ranks	Effect-Group Factor (−ve Value of D−R)	(D−R)
1	3	−1.339334033
2	5	−0.835096164
3	4	−0.756593956
4	2	−0.361083279

Finally, (D+R) and (D−R) values were plotted to obtain the causal diagram (Figure 3).

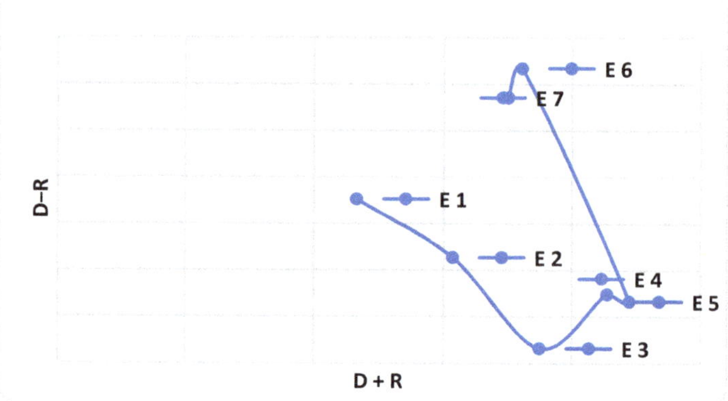

Figure 3. The Causal diagram.

Inferences

The values of (D+R) and (D−R) were calculated and shown in Table 8. Further, in Table 9, the values of (D+R) were ranked, where value delivery to stakeholders (5) got the highest value of (D+R), followed by transparency (4), authenticity (3), measurement and evaluation strategies (6), customer as co-creator (7), quality of content (2), distribution and promotion strategy (1). The factors with higher values of (D+R) show higher prominence with the system. Similarly, the positive and negative values of (D−R), as shown in Table 10 categorize the factors into cause or effect groups. The factors with a positive value of (D−R) are categorized into cause group factors. The factors with a negative value of (D−R) are categorized into effect group factors.

In this study, measurement and evaluation strategies (6); and the customer as co-creator (7) got the higher positive values of (D−R), which shows that these two factors have a high impact on other factors. But the (D+R) value of both the factors is low, which may be accounted for low levels of 'R'. The next factor with a positive but low value of (D−R) is distribution and promotion strategy (1), which shows that this factor doesn't impact other factors much. Also, this factor has the least (D+R) value, which shows that the factor does not carry much prominence with the system.

Further, in this study, four factors got categorized into effect group factors due to their negative (D−R) values, where authenticity (3), emerged with the highest value of negative (D−R), which shows that other factors are greatly impacting this factor. The high value of (D+R) also shows that this factor has high prominence with the system. The next two factors with high values of negative (D−R) are value delivery to stakeholders (5) and transparency (4). This shows that these two factors are also impacted by other factors significantly. Also, both these factors have the highest value of (D+R), making them significant and connected to the system. The factor with a low value of negative (D−R) is the quality of the content (2), and the same as a low value for (D+R) as well, which makes the factor getting moderately impacted by other factors.

Based on the outcomes from both the models, viz. TISM and DEMATEL, most factors emerged common in both the models, in terms of impact creating or dependency. Like, value delivery to stakeholders (5) emerged as the most dependant factor as per TISM model, and as per DEMATEL approach also this factor emerged with high prominence with the system, due to high value of (D+R) and got categorized into effect group, due to negative value of (D−R). Similarly, as per the TISM model, measurement and evaluation strategies (6); and the customer as co-creator (7) emerged as strong driving forces. As per the DEMATEL approach, these factors emerged as cause group factors due to positive values (D−R).

Further, authenticity (3) and transparency (4) emerged as dependant forces in the TISM model and DEMATEL approach as well both of them emerged as effect group factors with negative values of (D−R). But, the factors viz. distribution and promotion strategy (1); and quality of the content (2) emerged as driving powers in TISM, while in DEMATEL, distribution and promotion strategy (1) emerged as weak cause group factor, and quality of the content (2) emerged as weak effect group factor. This may be attributed to either less value of 'D' or 'R' associated with these factors.

5. Discussion and Practical Implications

The results obtained in the study highlight the relationship between the identified factors and the strength of their relationships as well. It emerged that branded content needs to promise value delivery to all the stakeholders, but the value delivery depends on several other associated factors. Customers should be promoted to be co-creator for branded content to enhance the credibility and acceptability of branded content. At the same time, the company needs to be vigilant in devising the measurement and evaluation strategies for the branded content. The measurement and evaluation strategies adopted by the company and decisions to involve customers as co-creator, directly and indirectly, impact the authenticity and transparency of the content. Customer trust needs to be created by developing quality content and appropriate distribution and promotion of the content. Due to this reason, companies take due care in adopting content distribution and promotion strategies.

This study emphasizes the strategic adoption and implementation of branded content as a marketing tool for the new normal (post-COVID-19 era). The marketers need to evolve continuously to keep the consumers engaged and attached to their brands. They need to cover up the limitations of traditional marketing, find ways to impact and convey their business values and digital branded content can be a tool for that. Digital content producers and advertisers need to integrate to frame powerful, thematic messages which can enhance organic viewership and brand preference [64]. In 2020, those companies who have focused on the content and its delivery could engage with customers more effectively. This pandemic has changed the habits and buying behaviour of consumers. While sitting and buying at home, the consumer looks for product reviews and details more minutely. As per a Forbes' study in 2016, the customers displayed 59% higher recalls than display advertising. The TISM tool applied in this study highlights that content distribution is very important for its success and in practical business life; this statement can be justified by looking at how 'Facebook' is giving success to the digital content distribution of companies. Apart from Facebook, marketers are exploring options of creating their advertisements and promoting through their apps like Apple news, google search and Snapchat, etc. [65]. This study emphasized the need for careful designing of quality branded content to ensure increased brand loyalty. Few studies in literature also substantiate this point whereby it is suggested to frame and deliver more informative content for high involvement product brands like a laptop; while for low involvement product brands like coffee, the content needs to be more attractive and attention seeker [9,27]. The way consumers perceive the usefulness and ease of use of content impact their attitude towards branded content, reflecting their purchase intentions further [26], which can benefit their brand if properly exploited by marketers establishment. Branded content can be a competitive tool for marketers in the coming days, wherein it can act as a bridge between the brand and the consumer relationship [66]. The younger generation mostly tends to escape from traditional advertising and feel connected with informative and entertaining promotions. There is always a quest for new content online which can be captured by branded content intelligently in the days to come.

6. Conclusions

Branded content is emerging as a subtle way of communicating about the brand with its users. The recent pandemic of COVID-19 has made consumers look towards online modes of buying, by choice or out of compulsion. Consumers now have to depend on online product or service reviews for making informed decisions. This is where the role of branded content pitches in. Marketers are innovating their ways to exploit the true worth of branded content as a robust marketing strategy. This study reflected the few factors that can enable branded content to be used as a robust marketing strategy. Authors have used the TISM and DEMATEL techniques to enable the strategic model framing of the identified enablers, to identify the appropriate way of approaching the enablers. This study outcome projected that branded content needs to provide value to different stakeholders, profit to the platform provider, increased sales to marketers, and genuine information to consumers. Value delivery again depends on the level of authenticity the content promises to the consumers. Marketing these days have become so vulnerable to mistrust and broken brand-consumer relationships. Stakeholders need transparency in content development and delivery. The quality of the branded content and the way it is delivered to the consumers make a huge difference to the success of this marketing strategy. The marketers, therefore, need to be very particular and focused on devising and implementing the measurement and evaluation strategies for the quality of the content and its delivery. The marketers need to draft the strategies depending on the type of product or service they are dealing with and the specific environment the brand is existing. Also, measuring the impact of branded content usage on the brand's sales is equally essential for marketers. Various marketers are exploring the options of including consumers in their process of branded content development to enhance the value of branded content. This study reflects the importance of strategically approaching the concept of branded content so that stakeholders get the real worth of this much-needed strategy.

7. Limitations and Future Research Directions

Like every research, this study also has limitations, including the sample size of respondents used to finalize factors. With more responses, a mixed-method approach could have been adopted for getting deeper insights. In the future, a similar study may be carried out by adopting techniques like factor analysis, structural equation modelling, and neural networking. A combination of Fuzzy-MICMAC, Fuzzy-TISM, and Fuzzy-DEMATEL can also be adopted to cover wider linkages and relationship strengths. The way different forms of branded content related to marketing strategy is worth research. Also, sector-specific or consumer-specific studies may be conducted wherein branded content is discussed as a marketing tool for specific products or services.

Author Contributions: Conceptualization, S.S.; Formal analysis, S.S.; Investigation, R.S.M.; Methodology, S.S.; Validation, R.S.M.; Writing—original draft, S.S.; Writing—review & editing, R.S.M. Both authors have read and agreed to the published version of the manuscript.

Funding: No funding was received for conducting this study.

Institutional Review Board Statement: Not applicable.

Informed Consent Statement: Not applicable.

Data Availability Statement: Not applicable.

Acknowledgments: Authors are thankful to the School of Management, The NorthCap University, Gurugram 122017, India and the Department of Food Engineering, National Institute of Food Technology Entrepreneurship and Management (NIFTEM), Kundli, Sonepat 131028, India, for providing the necessary facilities to support this research work.

Conflicts of Interest: The authors declare no conflict of interest.

Appendix A

Table A1. Interpretive Logic-Knowledge Base.

Sr. No.	Variable No./Paired Comparisons	Yes/No	Reason/Logic
E1 Distribution and Promotion Strategy			
1	E1-E2	NO	
2	E2-E1	NO	
3	E1-E3	YES	Choice of platform/channel used supports content authenticity
4	E3-E1	NO	
5	E1-E4	YES	Choice of platform/channel used supports transparency
6	E4-E1	NO	
7	E1-E5	YES	Choice of platform/channel used supports value delivery
8	E5-E1	NO	
9	E1-E6	NO	
10	E6-E1	YES	strong evaluation leads to better content marketing
11	E1-E7	NO	
12	E7-E1	YES	customer participates in decision making
E2 Quality of content			
13	E2-E3	YES	quality generates trust
14	E3-E2	NO	
15	E2-E4	YES	Transitive
16	E4-E2	NO	
17	E2-E5	YES	quality promises satisfaction
18	E5-E2	NO	
19	E2-E6	NO	
20	E6-E2	YES	stringent evaluation helps in quality content
21	E2-E7	NO	
22	E7-E2	YES	Customer participates in relevant content generation

Table A1. *Cont.*

Sr. No.	Variable No./Paired Comparisons	Yes/No	Reason/Logic
E3 Authenticity			
23	E3-E4	YES	Stakeholders will not hesitate in sharing authentic content
24	E4-E3	YES	more transparency ensures authentic content to be shared
25	E3-E5	YES	Satisfied stakeholders
26	E5-E3	NO	
27	E3-E6	NO	
28	E6-E3	YES	strong evaluation leads to better authenticity
29	E3-E7	NO	
30	E7-E3	YES	Transitive
E4 Transparency			
31	E4-E5	YES	Satisfied stakeholders
32	E5-E4	NO	
33	E4-E6	NO	
34	E6-E4	YES	strong evaluation leads to more transparency
35	E4-E7	NO	
36	E7-E4	YES	more involvement
E5 Value Delivery to stakeholders			
37	E5-E6	NO	
38	E6-E5	YES	Transitive
39	E5-E7	NO	
40	E7-E5	YES	Customer feel associated and gets value
E6 Measurement and evaluation strategies			
41	E6-E7	NO	
42	E7-E6	NO	

Appendix B

Table A2. Respondent details of Questionnaire.

Sr. No	Description	Details
1.	Total number of respondents	83
2.	Level of education	Undergraduate and above
3.	Occupation	Homemaker/student/employee

Few major Questions:

Q. 1. Do you prefer online shopping?

Q. 2. Have u started buying online, during the COVID-19 duration?

Q. 3. If No, what restricts you?

Q. 4. If yes, how is your experience?

Q. 5. While going for Online shopping, to how much extent do the content and reviews etc available on the website, influence your choice? Please rate on the given scale.

Q. 6. Please rate the below-mentioned variables on the scale of importance for the success of branded content for sales promotion:

Sr. No.	Variables	Very Important	Important	Can't Say	Not So Important	Completely of No Importance
i.	The strategy for distribution and promotion of the content					
ii.	The type of technologies used for creating the branded content					
iii.	The stage of the product for which content is created					
iv.	The content needs to be trustworthy (authenticate)					
v.	How much established the brand is					
vi.	How much transparent the content is					
vii.	Content quality needs to be good					
viii.	Content is paid one by the brand or is it free					
ix.	Value delivery to all the stakeholders (including you as a customer)					
x.	Whether the content relates well with traditional advertising					
xi.	Sufficient choices availability					
xii.	How much measurable the content is					
xiii.	Has the content evolved with time					
xiv.	You as a co-creator for the content					

References

1. Batra, D.; Singh, N. Marketing in the Era of Post-Covid Digitization—What Will a CMOs New Arsenal Be? 2020. Available online: https://www.businessinsider.in/advertising/brands/article/marketing-in-the-era-of-post-covid-digitization-what-will-a-cmos-new-arsenal-be/articleshow/78892267.cms (accessed on 24 January 2021).
2. Kamble, S.S.; Mor, R.S. Food supply chains and COVID-19: A way forward. *Agron. J.* **2021**, *113*, 2195–2197. [CrossRef]
3. Asmussen, B.; Wider, S.; Stevenson, N.; Whitehead, E. Defining Branded Content for the Digital Age: The Industry Experts' Views on Branded Content as a New Marketing Communications. Oxford Brookes University, 2016. Available online: www.thebcma.info/wp-content/uploads/2016/07/BCMA-Research-Report_FINAL.pdf (accessed on 20 April 2020).
4. Aguiar, A.C.W.C.; Steinhäuser, V. Branded Content-Strategic Marketing Tool and its Impact on Consumer Branded Content-Ferramenta Estratégica de Marketing e seuImpacto no Consumidor. *Braz. J. Mark. Res. Opin. Media* **2019**, *12*, 138–154.
5. Raju, A. Can reviewer reputation and webcare content affect perceived fairness? *J. Res. Interact. Mark.* **2019**, *13*, 464–476. [CrossRef]
6. Harp, E.A. Best Practices in Digital Branded Content for Generation Y: Developing Effective Campaigns in the New Era of Advertising. 2012. Available online: https://repositories.lib.utexas.edu/handle/2152/19954 (accessed on 12 June 2020).
7. Bhatia, V. Drivers and barriers of permission-based marketing. *J. Res. Interact. Mark.* **2020**, *14*, 51–70. [CrossRef]
8. Waqas, M.; Hamzah, Z.L.; Salleh, N.A.M. Customer experience with the branded content: A social media perspective. *Online Inf. Rev.* **2021**. [CrossRef]
9. Lou, C.; Xie, Q. Something social, something entertaining? How digital content marketing augments consumer experience and brand loyalty. *Int. J. Advert.* **2021**, *40*, 376–402. [CrossRef]
10. Pandey, N. Digital marketing strategies for firms in the post-covid-19 era: Insights and future directions. In *The New Normal Challenges of Managerial Business, Social and Ecological Systems in the Post Covid-19 Era*; Bloomsbury Prime: London, UK, 2021.
11. Graessley, S.; Horak, J.; Kovacova, M.; Valaskova, K.; Poliak, M. Consumer Attitudes and Behaviors in the Technology-Driven Sharing Economy: Motivations for Participating in Collaborative Consumption. *J. Self-Gov. Manag. Econ.* **2019**, *7*, 25. [CrossRef]
12. Hollowell, J.C.; Rowland, Z.; Kliestik, T.; Kliestikova, J.; Dengov, V.V. Customer Loyalty in the Sharing Economy Platforms: How Digital Personal Reputation and Feedback Systems Facilitate Interaction and Trust between Strangers. *J. Self-Gov. Manag. Econ.* **2019**, *7*, 13–18. [CrossRef]
13. Popescu, G.H.; Ciurlău, F.C. Making Decisions in Collaborative Consumption: Digital Trust and Reputation Systems in the Sharing Economy. *J. Self-Gov. Manag. Econ.* **2019**, *7*, 7–12.

14. Graham, K.W.; Wilder, K.M. Consumer-brand identity and online advertising message elaboration. *J. Res. Interact. Mark.* **2020**, *14*, 111–132. [CrossRef]

15. Mircică, N. Restoring Public Trust in Digital Platform Operations: Machine Learning Algorithmic Structuring of Social Media Content. *Rev. Contemp. Philos.* **2020**, *19*, 85–91. [CrossRef]

16. Drugău-Constantin, A.L. Is consumer cognition reducible to neurophysiological functioning? *Econ. Manag. Financ. Mark.* **2019**, *14*, 9–15.

17. Barreda, A.A.; Bilgihan, A.; Nusair, K.; Okumus, F. Generating brand awareness in Online Social Networks. *Comput. Hum. Behav.* **2015**, *50*, 600–609. [CrossRef]

18. Pop, R.A.; Săplăcan, Z.; Dabija, D.-C.; Alt, M.-A. The impact of social media influencers on travel decisions: The role of trust in consumer decision journey. *Curr. Issues Tour.* **2021**. [CrossRef]

19. Miliopoulou, G.-Z. Revisiting product classification to examine content marketing practices. *J. Res. Interact. Mark.* **2019**, *13*, 492–508. [CrossRef]

20. Vazquez, E.E. Effects of enduring involvement and perceived content vividness on digital engagement. *J. Res. Interact. Mark.* **2019**, *14*, 1–16. [CrossRef]

21. Lim, H.; Childs, M. Visual storytelling on Instagram: Branded photo narrative and the role of telepresence. *J. Res. Interact. Mark.* **2020**, *14*, 33–50. [CrossRef]

22. Lou, C.; Yuan, S. Influencer Marketing: How Message Value and Credibility Affect Consumer Trust of Branded Content on Social Media. *J. Interact. Advert.* **2019**, *19*, 58–73. [CrossRef]

23. Bratu, S. Can social media influencers shape corporate brand reputation? Online followers' trust, value creation, and purchase intentions. *Rev. Contemp. Philos.* **2019**, *18*, 157–163.

24. Holt, D. Branding in the age of social media. *Harv. Bus. Rev.* **2016**, *94*, 40–50.

25. Mwali, K. Branded Content Integration, Consumer Attitudes and Purchase Intent in South Africa. Master's Thesis, University of the Witwatersrand, Johannesburg, South Africa, 2016.

26. Sinthamrong, P.; Rompho, N. Factors affecting attitudes and purchase intentions toward branded content on webisodes. *J. Manag. Policy Pract.* **2015**, *16*, 64.

27. Lou, C.; Xie, Q.; Feng, Y.; Kim, W. Does non-hard-sell content work? Leveraging the value of branded content marketing in brand building. *J. Prod. Brand Manag.* **2019**, *28*, 773–786. [CrossRef]

28. Sabate, F.; Berbegal-Mirabent, J.; Cañabate, A.; Lebherz, P.R. Factors influencing popularity of branded content in Facebook fan pages. *Eur. Manag. J.* **2014**, *32*, 1001–1011. [CrossRef]

29. Moros, M. How do different digital channels of content marketing affect brand loyalty: A user-centric exploratory research. Bachelor's Thesis, Panteion University of Social and Political Sciences, Athens, Greece, 2021.

30. Ashley, C.; Tuten, T. Creative Strategies in Social Media Marketing: An Exploratory Study of Branded Social Content and Consumer Engagement. *Psychol. Mark.* **2015**, *32*, 15–27. [CrossRef]

31. Balan, C. Nike on Instagram: Themes of Branded Content and Their Engagement Power. In *CBU International Conference Proceedings*; CBU Research Institute: Prague, Czech Republic, 2017; Volume 5, pp. 13–18.

32. Feng, S.; Ots, M. Content marketing: A review of academic literature and future research directions. In Proceedings of the Emma Conference, Hamburg, Germany, 28–29 May 2015.

33. Eigenraam, A.W.; Eelen, J.; Verlegh, P.W. Let Me Entertain You? The Importance of Authenticity in Online Customer Engagement. *J. Interact. Mark.* **2021**, *54*, 53–68. [CrossRef]

34. Ikonen, P.; Luoma-aho, V.; Bowen, S.A. Transparency for sponsored content: Analysing codes of ethics in public relations, marketing, advertising and journalism. *Int. J. Strateg. Commun.* **2017**, *11*, 165–178. [CrossRef]

35. Carvajal, M.; Barinagarrementeria, I. The Creation of Branded Content Teams in Spanish News Organizations and Their Implications for Structures, Professional Roles and Ethics. *Digit. J.* **2021**. [CrossRef]

36. Edelson, L.; Chuang, J.; Franklin Fowler, E.; Franz, M.; Ridout, T.N. Universal Digital Ad Transparency. Available online: https://ssrn.com/abstract=3898214 (accessed on 25 August 2021).

37. Balint, A. Branded Reality: The Rise of Embedded Branding ('Branded Content'): Implications for the Cultural Public Sphere. Ph.D. Thesis, Goldsmiths, University of London, London, UK, 2016.

38. Kendall, L. Put a Price on Your Content: Measuring the Value of Content and Branded Content. Available online: http://kendallcopywriting.co.uk/wp/wp-content/uploads/2012/05/Put_a_price_on_your_content_measuring_the_value_of_content_and_branded_content.pdf (accessed on 12 June 2020).

39. Hajli, N.; Shanmugam, M.; Papagiannidis, S.; Zahay, D.; Richard, M.-O. Branding co-creation with members of online brand communities. *J. Bus. Res.* **2017**, *70*, 136–144. [CrossRef]

40. Koivisto, E.; Mattila, P. Extending the luxury experience to social media–User-Generated Content co-creation in a branded event. *J. Bus. Res.* **2018**, *117*, 570–578. [CrossRef]

41. Meilhan, D. Customer Value Co-Creation Behavior in the Online Platform Economy. *J. Self-Gov. Manag. Econ.* **2019**, *7*, 19–24. [CrossRef]

42. Baliga, A.J.; Chawla, V.; Sunder, M.V.; Kumar, R. Barriers to service recovery in B2B markets: A TISM approach in the context of IT-based services. *J. Bus. Ind. Mark.* **2021**, *36*, 1452–1473. [CrossRef]

43. Deshmukh, A.K.; Mohan, A. Analysis of Indian retail demand chain using total interpretive modeling. *J. Model. Manag.* **2017**, *12*, 322–348. [CrossRef]
44. Lianto, B.; Dachyar, M.; Soemardi, T.P. Modelling the continuous innovation capability enablers in Indonesia's manufacturing industry. *J. Model. Manag.* **2020**. [CrossRef]
45. Subadhra, R.; Suresh, M. Modelling of factors influencing brand sacralisation: A TISM approach. *Mater. Today Proc.* **2021**. [CrossRef]
46. Dehghan, A.; Mohammadkazemi, R.; Talebi, K.; Davari, A. Designing a Model of Empowerment for Small and Medium-Sized Businesses Knowledge-Based with a DEMATEL Approach. *Iran. J. Manag. Sci.* **2021**, *16*, 1–16.
47. Priya, S.S.; Priya, M.S.; Jain, V.; Dixit, S.K. An assessment of government measures in combatting COVID-19 using ISM and DEMATEL modelling. *Benchmark. Int. J.* **2021**. [CrossRef]
48. Shokouhyar, S.; Dehkhodaei, A.; Amiri, B. Toward customer-centric mobile phone reverse logistics: Using the DEMATEL approach and social media data. *Kybernetes* **2021**. [CrossRef]
49. Dahiya, S.; Panghal, A.; Sindhu, S.; Siwach, P. Organic food women entrepreneurs-TISM approach for challenges. *J. Enterp. Commun. People Places Glob. Econ.* **2021**, *15*, 114–136. [CrossRef]
50. Sushil. Multi-criteria valuation of flexibility initiatives using integrated TISM–IRP with a big data framework. *Prod. Plan. Control.* **2017**, *28*, 999–1010. [CrossRef]
51. Dwivedi, A.; Madaan, J. A hybrid approach for modeling the key performance indicators of information facilitated product recovery system. *J. Model. Manag.* **2020**, *15*, 933–965. [CrossRef]
52. Jena, J.; Sidharth, S.; Thakur, L.S.; Pathak, D.K.; Pandey, V. Total Interpretive Structural Modeling (TISM): Approach and application. *J. Adv. Manag. Res.* **2017**, *14*, 162–181. [CrossRef]
53. Mor, R.S.; Bhardwaj, A.; Singh, S. Benchmarking the interactions among barriers in Dairy supply chain: An ISM approach. *Int. J. Qual. Res.* **2018**, *12*, 385–404. [CrossRef]
54. Lamba, K.; Singh, S.P. Modeling big data enablers for operations and supply chain management. *Int. J. Logist. Manag.* **2018**, *29*, 629–658. [CrossRef]
55. Chen, C.A. Using DEMATEL method for medical tourism development in Taiwan. *Am. J. Tour. Res.* **2012**, *1*, 26–32.
56. Chiu, Y.-J.; Chen, H.-C.; Tzeng, G.-H.; Shyu, J.Z. Marketing strategy based on customer behaviour for the LCD-TV. *Int. J. Manag. Decis. Mak.* **2006**, *7*, 143. [CrossRef]
57. Lin, S.M. Marketing mix (7P) and performance assessment of western fast food industry in Taiwan: An application by associating DEMATEL and ANP. *Afr. J. Bus. Manag.* **2011**, *5*, 10634–10644.
58. Susanty, A.; Puspitasari, N.B.; Prastawa, H.; Renaldi, S.V. Exploring the best policy scenario plan for the dairy supply chain: A DEMATEL approach. *J. Model. Manag.* **2021**, *16*, 240–266. [CrossRef]
59. Wang, Y.-L.; Tzeng, G.-H. Brand marketing for creating brand value based on a MCDM model combining DEMATEL with ANP and VIKOR methods. *Expert Syst. Appl.* **2012**, *39*, 5600–5615. [CrossRef]
60. Maheshwari, P.; Seth, N.; Gupta, A.K. An interpretive structural modeling approach to advertisement effectiveness in the Indian mobile phone industry. *J. Model. Manag.* **2018**, *13*, 190–210. [CrossRef]
61. Medalla, M.E.; Yamagishi, K.; Tiu, A.M.; Tanaid, R.A.; Abellana, D.P.M.; Caballes, S.A.; Jabilles, E.M.; Himang, C.; Bongo, M.; Ocampo, L. Modeling the hierarchical structure of secondhand clothing buying behaviour antecedents of millennials. *J. Model. Manag.* **2020**, *15*, 1679–1708. [CrossRef]
62. Ghatak, R.R. Barriers analysis for customer resource contribution in value co-creation for service industry using interpretive structural modeling. *J. Model. Manag.* **2020**, *15*, 1137–1166. [CrossRef]
63. Sheoran, M.; Kumar, D. Modelling the enablers of sustainable consumer behaviour towards electronic products. *J. Model. Manag.* **2020**, *15*, 1543–1565. [CrossRef]
64. Yadav, M. Brand Integrated Content on Digital Media An Untapped Opportunity for Marketers in the Post COVID World. Available online: http://www.businessworld.in/article/Brand-Integrated-Content-on-Digital-Media-an-untapped-opportunity-for-marketers-in-the-post-COVID-world-/01-09-2020-315460/ (accessed on 19 December 2020).
65. Berry, E. The Changing Face of Branded Content. 2018. Available online: https://www.forbes.com/sites/forbesagencycouncil/2018/05/18/the-changing-face-of-branded-content/?sh=6131a40c4c2d (accessed on 15 October 2020).
66. Bezbaruah, S.; Trivedi, J. Branded Content: A Bridge Building Gen Z's Consumer–Brand Relationship. *Vision* **2020**, *24*, 300–309. [CrossRef]

systems

Article

Complexity Economics in a Time of Crisis: Heterogeneous Agents, Interconnections, and Contagion

Michael S. Harré [1],*, Aleksey Eremenko [2], Kirill Glavatskiy [1], Michael Hopmere [3], Leonardo Pinheiro [2], Simon Watson [1] and Lynn Crawford [3]

[1] Complex Systems Research Group, Faculty of Engineering, The University of Sydney, Camperdown 2006, Australia; kirill.glavatskiy@sydney.edu.au (K.G.); swat8088@uni.sydney.edu.au (S.W.)
[2] Independent Researcher, Sydney 201101, Australia; aleksey_eremenko@hotmail.com (A.E.); leosantospinheiro@gmail.com (L.P.)
[3] School of Project Management, Faculty of Engineering, The University of Sydney, Camperdown 2006, Australia; mhop9461@uni.sydney.edu.au (M.H.); lynn.crawford@sydney.edu.au (L.C.)
* Correspondence: michael.harre@sydney.edu.au

Abstract: In this article, we consider a variety of different mechanisms through which crises such as COVID-19 can propagate from the micro-economic behaviour of individual agents through to an economy's aggregate dynamics and subsequently spill over into the global economy. Our central theme is one of changes in the behaviour of heterogeneous agents, agents who differ in terms of some measure of size, wealth, connectivity, or behaviour, in different parts of an economy. These are illustrated through a variety of case studies, from individuals and households with budgetary constraints, to financial markets, to companies composed of thousands of small projects, to companies that implement single multi-billion dollar projects. In each case, we emphasise the role of data or theoretical models and place them in the context of measuring their inter-connectivity and emergent dynamics. Some of these are simple models that need to be 'dressed' in socio-economic data to be used for policy-making, and we give an example of how to do this with housing markets, while others are more similar to archaeological evidence; they provide hints about the bigger picture but have yet to be unified with other results. The result is only an outline of what is possible but it shows that we are drawing closer to an integrated set of concepts, principles, and models. In the final section, we emphasise the potential as well as the limitations and what the future of these methods hold for economics.

Keywords: complexity economics; economic crisis; COVID-19; agent-based model; information theory; global value chains; megaprojects; housing markets; economic networks

Citation: Harré, M.S.; Eremenko, A.; Glavatskiy, K.; Hopmere, M.; Pinheiro, L.; Watson, S.; Crawford, L. Complexity Economics in a Time of Crisis: Heterogeneous Agents, Interconnections, and Contagion. *Systems* **2021**, *9*, 73. https://doi.org/10.3390/systems9040073

Academic Editors: Oz Sahin, Alberto De Marco and Paolo Visconti

Received: 14 July 2021
Accepted: 9 October 2021
Published: 15 October 2021

Publisher's Note: MDPI stays neutral with regard to jurisdictional claims in published maps and institutional affiliations.

1. Introduction

1.1. The Economics of Heterogeneity and Interconnections

Crises that disturb the economic status quo have a ripple effect that can reverberate through markets and economies around the world. The effect a crisis has on any individual entity depends on the characteristics of the entity and the nature of its connections with the rest of the economy, and this is one of the areas that complexity economics (CE) has been able to contribute to [1]. CE has its origins at the Santa Fe Institute in the mid-1980s when economists, computer scientists, and physicists came together to foster an interdisciplinary approach to addressing economic problems. Since then, a number of groups and institutions have sprung up around the world such as Oxford University's Institute for New Economic Thinking (INET) headed by J-D Farmer, Harvard's Atlas of Economic Complexity (AEC[1] [2]), and MIT's Observatory of Economic Complexity (OEC[2]), these having developed through the work of Hausmann, Hidalgo and colleagues [3,4]. Recently, some central banks, such as the Bank of England [5] and the Canadian Central Bank [6], have begun to explore CE methods, such as agent-based models (ABMs), for policy-making and in their approaches to modelling market dynamics. This is an approach that has been growing in its sophistication and accuracy with a paper by Poledna and colleagues [7] having recently won 1st prize in the Complexity and Macroeconomics Competition held by *Rebuilding Macroeconomics* for producing an ABM that is comparable in forecasting ability to traditional DSGE (Dynamic Stochastic General Equilibrium) models.

CE has sometimes been critiqued for not being a single theory or a unified approach to economics [8]. This is in part because in practice it is an ecology of ideas, analogies, and methods combined with large amounts of domain-specific data that are used to address particular problems, freely borrowing from other fields in order to do so. However, this also belies a technical consistency in the approach CE adopts. In particular it focuses on the formal strengths of models that have often been validated in other fields and then applies the appropriate economic framework around the analysis, thus allowing models to develop independently of both economic ideology and the other fields that inspired the initial analogies. For example, 'spin models' [9,10], binary state models that represent agents as, for example, buyers and sellers, have been used extensively as simple models by CE theorists to develop their intuition for the micro-economic agent-to-agent interactions that result in the emergence of nonlinear macro-economic dynamics. The phase transitions we see in spin models then help frame our thinking of the non-linearities of complex market behaviour such as the two-phase dynamics discovered by Plerou et al. [11]. This has also been used extensively in the work of Brock and Durlauf [12] and Aoki [13], for example, in understanding the interactions between agents and the emergence of multiple equilibrium.

These models first appeared in the field of physics which remains a significant source of inspiration for CE [14,15], as does the mathematics of network theory [10,16] and evolution [17,18], ideas that have been brought together in work on evolutionary game theory on networks in order to understand the emergence of cooperation [19] and spreading dynamics over networks [20]. On the topic of evolution in economics, Brian Arthur, who coined the termed 'complexity economics', has written [21]: ... *because complexity economics looks at how structures form or solutions come to be 'selected', it connects robustly with the dynamics of evolutionary economics.* Evolutionary economics also explicitly accounts for the path dependencies of outcomes and the heterogeneity of agents, as Volmir writes in his review [22] of Nelson et al's book [23]:

> ... technological trajectories are a cumulative process of searching for "new ways to do things", providing the reader with a framework to explain emerging behaviors such as lock-ins, 'anti-commons' problems... Since the 1960s, innovations began to be viewed as multi-interactive phenomenon, which entails a cumulative process between different agents and institutions, a fact ignored by standard economics... Once the cumulative process is understood, it is impossible to deny that there are differences in the ability of distinct firms to accumulate knowledge.

This combination of heterogeneity, path dependency, interactivity, and innovation are all hallmarks of the 'world view' of CE.

With these concepts in mind, theory and empirical study in economics have been moving into an era of networks and heterogeneous agents, and along with this progression comes a growing awareness of the systemic risks of a highly connected society. Take, for example, the network of international trade relations that have been a lynch pin of modern trade development recently reviewed by Carrère et al. [24]. In this domain, a central model has been the gravity model of trade [25], and in its simplest form is conceptualised similarly to that of gravity in physics. In physics, the attractive forces F between two objects of masses M_1 and M_2 is proportional to the product of their masses divided by the square of the radial distance r^2 between them:

$$F = G\frac{M_1 M_2}{r^2} \tag{1}$$

where G is the universal gravitational constant. The gravity model of international trade has a very similar form where the trade flow $T_{i,j}$ between two economies with gross domestic products of P_i and P_j and the geographical, political, social, or some other measure of separation is represented by $D_{i,j}$, is expressed by the relation

$$T_{i,j} = C\frac{P_iP_j}{D_{i,j}} \tag{2}$$

where C is a constant of trade. See Anderson [26] for the theoretical foundations and extensions. This formulation, first framed in 1954 [25], lends itself naturally to a network analysis in which each $T_{i,j} = T_{j,i}$ is the symmetrical weighted link in a network connecting nodes of size P_i and P_j. For a fixed distance, as the sizes of the economies grow the flow of trade between them increases, but if the distances between these same economies were increased the trade flows would decrease in inverse proportion. The analysis of these networks has been further extended to the micro-economic level through the work of Bergstrand [27], for example, providing a much more granular conceptualisation of trade in which the gravity model is a reduced form of a more sophisticated partial equilibrium model of trade.

The network model of trade is more general than the gravity model though. For example, Rauch has studied the social networks of trade [28] and the relationships between networks and markets [29]. In a similar study, Chaney [30] looked at the network structure of trade where firms can only export into markets in which they have contact and acquire new contacts both at random as well as through their network of existing contacts, thereby introducing an element of randomness to network formation. The specific varieties of what is traded over these networks have also been considered in some detail, with Dalin et al. [31] looking at the trade in 'virtual water', the amount of water needed to produce food, and the global trade in arms studied by Arkerman and Seim [32]. Alongside this appreciation of the role of network topology and flows of trade is a further appreciation of the heterogeneity of the nodes themselves, i.e., the highly varied characteristics of the countries, companies, and individuals that are the linked agents.

The recent growth of research in this area has been stimulated, at least in part, by the growth in international trade and the use of Free Trade Agreements between a large number of economies and consequently a need to better understand how this has shifted regional economies. This is because although there had been no global free trade agreements since 1994 [33], whereas the number of regional or bilateral trade agreements between countries (e.g., NAFTA, EU, and APEC) grew from 50 in 1990 to more than 280 in 2017 [34]. This has led to a deeper interest in systemic risks such as the fragility of supply chains [35], the interaction between trade networks, trade wars, and firm value [36], as well as trade related climate change [37]. However, this has been in parallel with an enormous growth in the study of 'complex systems' through the lens of network analysis, a field in which recent research began with Watts and Strogatz in 1998 [38] and Barabási et al. in 1999 [39], and it has played a core role in many fields over the last twenty years. This has particularly been the case in economic research where the formal methods of network science have been used by non-economists, often but not always physicists, to study the abstract properties of trade networks [40–42]. While the recent connection between physics models and economic data has not always been a harmonious one, it has been productive in certain fields of economics with several articles by prominent researchers on both sides of the debate having voiced strong opinions on the success or otherwise of these methods [43–45].

These debates have been had at the macro-economic level as well as at the market and micro-economic levels, all of which has been a part of a steady revolution in economic thinking over the last 30 years. Some of the earliest work using simulations or a 'complex systems' approach [46] includes the work of Brian Arthur and colleagues on the simulation of financial markets [47] and other models of collective economic behaviour [48] where traditional assumptions, such as equilibrium or rational choice, are relaxed in order to study the evolutionary dynamics of markets and under what circumstances an equilibrium state might naturally come about with these assumptions relaxed [49]. Other models, such as bifurcation models in which a form of dynamic equilibrium is presumed, are suited to partial equilibrium analysis in the sense that it is the non-equilibrium transition between alternative stable states that is most interesting, see Rosser Jr.'s review of economic Catastrophe Theory for example [50].

Within the context of complexity economics, this article reviews several recent research directions at multiple different levels of analysis as well as some of the work that we have carried out in recent years. This includes our work on applied network theory [20,51–54], bifurcations and systemic risks [55–60], agent-based modelling of economic markets [61–64], the theoretical limits of 'rationality' and strategic choice [20,65–69], and how information theory can be used to understand the dynamics of these systems [70–77]. The purpose then is to place this research in the context of the work being carried out in other groups around the world. We hope that further developments of these areas will ultimately lead to larger models that can be used to better understand the macro-economic response of an economy to global shocks during a crisis such as COVID-19 or the Global Financial Crisis (GFC). In the following subsections, we introduce the central themes of this work that are the basis of the sections in the main body of this article.

1.2. The Household Level: Theory and Simulation

As we write in mid-2021 the pandemic continues to push economies around the world into lockdowns where social distancing measures are put in place, restricting our freedom of movement as well as the ability of the economy to function properly. At the level of households, the first two sections investigate if the financial distress caused by the pandemic could cause a new period of stress in economic markets such as the one observed during the GFC. This topic has two parts: the first is a stylised agent-based simulation without any real-world data, and the second is a more realistic simulation using real data from the Greater Sydney housing market. The purpose of these case studies is to illustrate the strengths and weaknesses of the two approaches as well as the relationship between them.

In Section 2, we implement the modified diffusion model with financial constraints first proposed by Gallegati et al. [78] in order to model the 'period of financial distress' prior to a market decline for markets in general and later used as a model of housing markets specifically [79]. At the individual agent level such a period might occur when the agent, such as a firm or a household, is faced with the not yet realised but highly probable chance of not being able to meet their financial obligations [80]. In the broader sense of a whole market this can occur because a subgroup of agents have wealth constraints that limit their ability to buy assets outright and so they need to borrow to buy assets that then, through the evolution of the market price, become undervalued and as the assets are then distressed some over-leveraged agents need to sell, pushing prices down even further than fundamentals suggest is the equilibrium price. The effect of leverage on market stability has been extensively studied in housing markets [81,82] and agent-based models of financial markets [83,84] and in the work of Kindleberger [85], periods of financial distress are a general pattern in many bubbles and their subsequent crashes throughout the last several centuries. In Kindleberger's words [85] (p. 11):

> Then an event—perhaps a change in government policy, an unexplained failure of a firm previously thought to have been successful—occurs that leads to a pause in the increase in asset prices. Soon, some of the investors who had financed most of their purchases with borrowed money become distress sellers of the real estate or the stocks because the interest payments on the money borrowed to finance their purchases are larger than the investment income on the assets. The prices of these assets decline below their purchase price and now the buyers are 'under water'—the amount owed on the money borrowed to finance the purchase of these assets is larger than their current market value.

In Section 2, we reproduce the model of Gallegati et al. [78] to illustrate this period of financial distress.

In Section 3, we move from these theoretical considerations towards the more applied level of the Australian government's response to the COVID-19 'event', to use Kindleberger's terminology, and its impact on the housing market. In response to the pandemic, the government moved to close borders [86] that reduced the influx of temporary residents (e.g., students and short-term workers), resulting in a decrease in the demand for rental properties and the corresponding decline of the rental prices [87] which has impacted the income of property investors. Second, decreasing the cash rate by the Reserve Bank of Australia [88] has increased the incentive for mortgage borrowing among households that are otherwise stressed due to COVID-19, which has resulted in an increased demand in housing and a corresponding increase in prices [89]. Third, the government's 'JobKeeper' and 'JobSeeker' payment schemes [90,91] intended to support households' individual budgets and to stimulate their consumption activity created an auxiliary source of income for households, which has arguably altered their budgeting incentives.

These government policy-driven macro-economic factors have combined with the micro-economic effects of reduced spending for holidays due to travel restrictions and other changes in household behaviour such as reductions in food wastage (as reported in other countries during COVID-19 [92]) and so impacting household expenditure, the consequences of which is household savings of \$100 billion during 2020 [93].[3] This has had the consequential effect of increasing pressure on the housing market in 2021 as the extra savings has fuelled further interest in house buying across Australian markets. As the Australian Bureau of Statistics reported [94]:

> Increased housing market activity was driven by an expansive monetary policy and support through government policies such as Homebuilder and other state specific initiatives, as well as pent up demand (due to lower activity during the June quarter [2020] COVID-19 lockdown period). As auctions and open home inspections picked up in September quarter (with the easing of social distancing measures), greater demand than there was housing stock on the market saw property prices rebound

These factors have contributed to an already highly valued Australian housing market [62,63] and will likely continue to contribute to wealth and housing inequality into the future. We emphasise this combination of government policy and micro-economic factors because we want to illustrate how emergent consequences arise from changes in individual behaviours during a crisis, something that needs to be explicitly modelled because there is usually, as in the case of the GFC and COVID-19, no previous macro-economic data on which to base sound judgement, so estimates of the impact of individual behaviour need to be used, and how these behaviours drive the macro-economic dynamics that policy-makers want to manage.

With this Australian-specific perspective in mind, in Section 3 we compare the methods of Section 2 with those of a model of the Greater Sydney housing market to illustrate the strengths of a realistic model that uses high-resolution market and socio-economic data in which household constraints are varied to reflect key aspects of the COVID-19 crisis. We show that the effect of COVID-19 on Sydney house prices is the opposite to that of a bubble–crash dynamic; house prices increase significantly in the model and this has been observed in recent price movements across Sydney and other Australian capital cities, lending significant credence to the use of agent-based models to simulate out-of-sample dynamics during a crisis.

1.3. Financial Markets and Systemic Risks

Financial market crises are a common topic of study for researchers outside of mainstream economics. This is due in part to the extensive amount of data available which allows financial markets to be analysed very well with the tools of computer science and physics that were initially developed for studying large stochastic systems with interacting elements. Some of the earliest work in this area that continues to drive research is in the study of the so-called stylised facts of financial markets, such as fat tails and clustered volatility [95]. One of the earliest debates was over the best model to use for market fluctuations: cascading turbulence or a truncated Lévy flight [14,15,96,97]. This led to a long and fruitful series of investigations into the dynamics of the univariate time series of financial market indices [98–101], with recent contributions from our group in this area as well [102–105].

At a more granular level, the analysis of markets can be seen as an interaction between prices that, to at least some extent, contribute to the (co-)movement of other prices, inducing a dynamical asset network that can be studied for its stability properties. This type of analysis can, for example, extract market sectors from price movements by examining the largest eigenvalues of the market's cross-correlation matrix [106] as well as using random matrix theory to distinguish between random and non-random correlations [107], methods that were originally motivated by models in physics. As an approach to understanding crises, such as the Black Monday crash of 1987, and optimal portfolio selection, Onella and colleagues applied correlation-based network analyses in order to understand market risks [108–111]. Our group has extended these methods to information theoretical methods in order to study the nonlinear properties of markets and other types of nonlinear dynamics [52,55,70,112]. Extending these methods by using information transfers between equities such as Granger causality [113] or transfer entropy [71,114] results in fundamentally different networks of relationships between equities [70] and consequently different risk profiles depending on the different measures of relationship used.

In order to study some recent market events that have been particularly turbulent and have yet to be studied in detail, in Section 4 we use transfer entropy (see, for example, in [70] and [Chapter 6] in [71]) to infer a temporal flow of information between equities during the periods covering three market events: the US Federal bail out decision of 2008, the Flash Crash of May 2010, and the COVID-19 crash of 2020, together with three other random control dates. For each of these events, we use tick-by-tick financial data that allow us to study the day before the event, the day of the event, and the day after the event with three thirty-minute periods used to compute transfer entropy of the Dow Jones Index. This gives us considerable fine grained insight into the micro-evolution of information flows through the market and their relationship to overall market dynamics.

1.4. Trade Networks: Internal and External Trade in Value Added

As mentioned in the first section above, trade networks are vital to the economic development and prosperity of a country's economy. Since the earlier work on gravity models, work has developed extensively in understanding the relationship between trade within a country's economy and trade between different economies. Developments such as the Observatory of Economic Complexity [115] (https://oec.world, accessed on 13 October 2021) have taken significant amounts of trade data and converted it into country-specific trade network analyses that can be drilled down into the sub-market sectors as well as, in many instances, the distinct economic regions within a country. Research stemming from this work has established important links between these trade networks and the specifics of income inequality [116], the environment [117], and employment [118], all issues that are of central concern to the sustainable growth and development of a country.

These networks have developed much more slowly over time than other networked aspects of the economy such as financial markets or housing markets, and their disruption and subsequent recovery might also be expected to be somewhat slower. However, even short-term shocks to trade networks can have a significant and long-lasting impact on trade links such as agri-food trade networks [119,120] and other commodities [121,122] that are heavily traded as physical goods across the globe. These have been studied in detail for previous crises, for example the role of the inter-bank network of debt during the GFC [123]. In Section 5, we look at the intra- and inter-economic trade data beginning at the industry sector level to examine patterns in the trade of goods and services between market sectors. In working to understand the long-term shocks caused by COVID-19, we can use analyses of this type to form a better picture of the complexity and interrelationships of sub-market trading and its implications of policies.

1.5. Business Sector Analysis

In Section 6, we discuss the impact of global crises on businesses through the lens of projects and their role as vehicles for economic transformation. At an operational level, projects are a useful framework within which organisations can plan and control the delivery of products and services that generate income or otherwise benefit businesses and their customers [124]. As is the case for other elements of economic development, the effect that a crisis has on businesses or individual projects depends on their individual characteristics and the nature of their connections with the rest of the economy. Large projects and large portfolios of related smaller projects can have an out-sized impact on economic development and their stalling or failure during a crisis has a consequential knock-on effect for the rest of the economy. Using data reported during the COVID-19 pandemic, we show how projects are strongly connected with, and have been impacted by, their respective economies, with a significant number of projects being cancelled or suspended.

In particular, we look to a class of projects known as 'megaprojects' that are commonly used to deliver very large, complex, and costly outputs such as infrastructure, water, energy, and mining ventures [125–127]. Their extreme scale is reflective of the functional complexity of megaprojects which are themselves often initiated to facilitate the productive efficiency and delivery of many other goods and services. In other words, large, complex projects are often singular economic exercises around which other economic developments organise themselves, providing support to downstream development projects in multiple industrial sectors.

This is most apparent in multi-billion dollar projects such as energy or road projects that facilitate further development throughout the economy. For example, Olds [128] has examined urban mega-projects on the Pacific rim (Vancouver, Yokohama, and Shanghai) and the relationship between local economic development and globalisation. Similarly, Zekovic et al. [129] has looked at megaprojects in the context of urban planning and development. Most telling of all though is the enormous amount of infrastructure that is required to support GDP growth in the coming decades and the role of megaprojects in this development. In a 2017 article, Söderlund et al. [130] wrote:

> One reason for such acceleration in megaprojects can be gleaned from the projections of infrastructure to meet the world's ever-increasing needs for economic growth and improvements. McKinsey (Garemo, Matzinger, & Palter, 2015) estimates that the world needs to spend about US\$57 trillion on infrastructure by 2030 to keep up with the expected GDP growth. The Organisation for Economic Co-operation and Development (OECD) estimates that 'global infrastructure investment needs of US\$6.3 trillion per year over the period of 2016–2030 to support growth and development', which exceeds the figure proposed by McKinsey.

Due to their extreme scale and impact, megaprojects often play a pivotal role in the shaping of an economy. However, the inverse may also be argued, i.e., where there is a disruption like that caused by a global shock such as COVID-19 this can precipitate the cancellation or pausing of megaprojects as their funding bodies reassign previously budgeted capital to other more pressing needs and the resultant loss or delay of innovation and value. The net effect is to push back the infrastructure necessary to support the economic growth of the next decade.

As some countries begin to emerge from the grip of the pandemic, there are moves to spur economic recovery by initiating large infrastructure projects. Starting (or restarting) significant numbers of projects over a short period will likely stress delivering organisations as they seek to rapidly revive projects and recover the work force at the same time as competing for newly announced infrastructure projects. This restart may create shortages of raw materials as global supply chains respond to rapidly increasing demand and competition as skilled resources, stressing already compromised global value chains such as those described in Section 5.

1.6. The Structure of the Article

In the sub-sections above, we have given a brief overview of some of the recent literature as well as a review of the tools that are used in what follows in this article. The progression of the sections from Section 2 through Section 5 are in approximate order of increasing degrees of coarse-graining. Section 2 is purely theoretical and every agent is completely described at the discretion of the modeller, however this level of information comes at the price of lacking in real-world precision. Section 3 is again a simulation, but it is populated with realistically calibrated agents and an economic context that is based on the real world while also being expected to faithfully reproduce empirical observations of market behaviour. Section 4 has no explicit agents or models of interactions, instead it empirically examines the interactions between agents that can be inferred using multivariate time series from financial markets. Section 5 looks at an even further aggregated level, starting from data of entire industry sectors and looking at the trade in value added between sectors within an economy as well as between economies. Although each of these layers interact with each other to produce multiply layered networks of interactions, we do not yet integrate them in a unified model; this is left for further development and is an open research effort in this field. Finally, in Section 6, we discuss the unique aspects of project economics and their role in development, both during and post an economic crisis.

Each section should be seen as a distinct and relatively independent case study that illustrates methodologies or principles in action, without going into a great deal of detail (references are provided to work describing the relevant details). In particular, we have used the methods described above to illustrate various strengths and weaknesses of particular ideas from CE applied to the analysis of an economy, usually with an Australian perspective for concreteness. This is not intended to be a complete analysis of an economy, for example, we have omitted central banks, commercial banks, and other key institutions and markets. Nor should these be seen as an integrated approach to modelling an economy because we have yet to make clear the connections between the different elements we will describe. Instead, the intention is to illustrate the multiple directions of research that are being pursued and that are now being drawn together to contribute to a unified "whole of system" approach to modelling large scale economic phenomena. In the meantime, it is hoped that this article gives some insight into what is presently being developed and what the future holds for the field.

2. Periods of Financial Distress in an Agent Based Model

As mentioned in Section 1, in the work of Kindleberger [85] the initial cause of financial distress can be the action of a few agents that by some means come to believe that they are in stress or a bubble and that, once they have acted on this belief, the rest of the market comes to a similar realisation and that many of them may need to sell in order to manage their financial position. What initiates a market sell-off is a matter of ongoing debate so to make our ideas concrete in the form of a relevant simulation we follow the work of Gallegati et al. [78] in order to simulate an endogenous market crises. The purpose in using this model is that it includes agent level (household) financial constraints as a primary contributor to an asset market crisis, where the asset might be equities, houses, or something else. These constraints are related to the costs of buying and selling the asset, which are implicitly relative to the total household budget, if either the household budget changes or the transaction costs change for a large enough portion of the market then a market-wide period of financial distress may result, and it is these shifting household constraints that we will relate to the COVID-19 crisis in Section 3.

2.1. The Theoretical Framework

In [78], there is a population of buyers and sellers facing a binary choice problem for a single risky asset. The simulation runs for a number of steps indexed by t and at each step an agent i chooses a strategy $w_{i,t} \in \{-1, 1\}$ where -1 represents selling and $+1$ represents buying. The asset's underlying price dynamic is given by

$$p_t = p_{t-1} + kw_t + \sigma z_t \tag{3}$$

Here, the price evolution is a sequence of $n + 1$ values $\{p_0 \ldots p_n\}$ with a given p_0, z_t is a Weiner process (noise), and σ is the strength of the noise. Excess demand is the average of all agents' choices at time t where the mean strategy is $w_t = \langle w_{it} \rangle$ and therefore kw_t is the k-weighted influence that excess demand has on price (set to $k = 0.4$ in the original paper). The expected excess demand at t is assumed to be $w_t^e = w_{t-1}$. The utility for each agent i is given by

$$U_{it} = (\bar{p}_t - p_{t-1})w_{it} + Jw_{it}w_t^e + \epsilon_{it} \tag{4}$$

in which $\bar{p}_t$ is a single stochastic adaptive learning process for all agents:

$$\bar{p}_t = \bar{p}_{t-1} - \rho(\bar{p}_t - p_t) + \sigma_1 z_{1,t} \tag{5}$$

and $\rho \in [0, 1]$ controls the adaptation speed. The strength of the agents' interaction with one another is controlled by the parameter J, as it is through the $Jw_{it}w_t^e$ term that individual choices are connected to all other choices via a mean market estimate of excess demand. The agents make their decisions based on the expected benefit of trading using both the recent observable price changes and the relative excess demand weighted by a herding factor (see Equation (4)). The decision-making process each agent uses is based on a value function $V_{i,t}$:

$$V_{i,t} = \begin{cases} U_{it} & \text{if } W_{i,t-1} > \theta W_{i,0} \\ -\infty & \text{if } W_{i,t-1} \leq \theta W_{i,0} \end{cases} \tag{6}$$

where $W_{i,t}$ is the wealth of agent i at time t, $W_{i,0}$ is the initial wealth of each agent and θ is some real valued proportion of the initial wealth such that, if the wealth falls below a threshold value of an agent's initial wealth then they will not trade. Then, a choice $w_{it} \in \{-1, 1\}$ is made according to the probability:

$$P(w_{it}) = P(V_{it}(w_{it}) > V_{it}(-w_{it})) \tag{7}$$

This trading model follows the social interaction approach of Brock and Durlauf [12]. The full simulation can be summarized as carrying out the following steps for a total number of T iterations:

1. Compute agent's decisions w_{it} using Equations (4), 6, and 7,
2. Asset prices are updated using Equation (3),
3. Agents realize profit/losses and update their wealth,
4. Agents compute a new expected price.

For a complete description and background of the algorithm see the original paper [78].

2.2. Simulation Results

By reproducing the original model we are able to achieve similar results to the original article, and we note the following:

- A bubble characterized by a PFD is produced only when transaction costs are sufficiently high, see Figure 1. In the absence of high transaction costs no crashes are observed in the simulations.
- High financial costs cause a pattern of crashes in Figure 2). As a hypothesis for the cause of financial distress, high transaction costs have the drawback of causing repeatable patterns that are may not be realistic.

- The evolution of the wealth and the distribution of the agents explain the bubble (Figure 3). We can see two densities of wealth that correspond to the beginning of the simulation and right before the crash. It demonstrates that financial distress seems to be correlated with the occurrence of shocks.
- Changes in the herding factor, J, affect the amplitude of the bubble, making social interaction an important component of how financial contagion spreads and how the shock ultimately unfolds into a crisis.

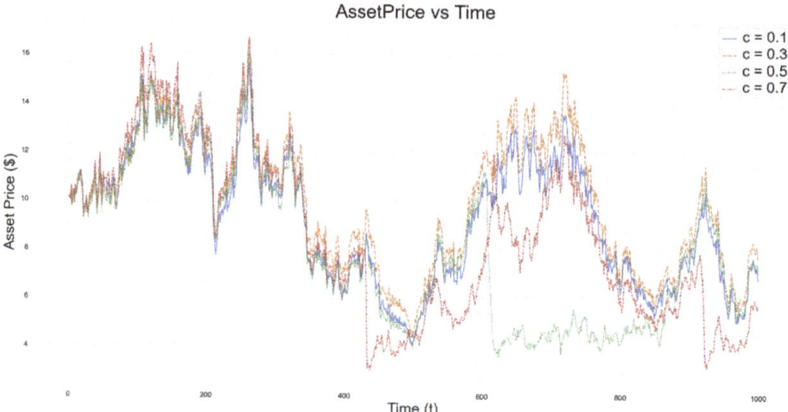

Figure 1. Comparison of multiple values for transaction costs. For $c = 0.5$ there is a crash at $t \approx 900$. and for $c = 0.7$ multiple crashes occur. There are no crashes for $c < 0.5$.

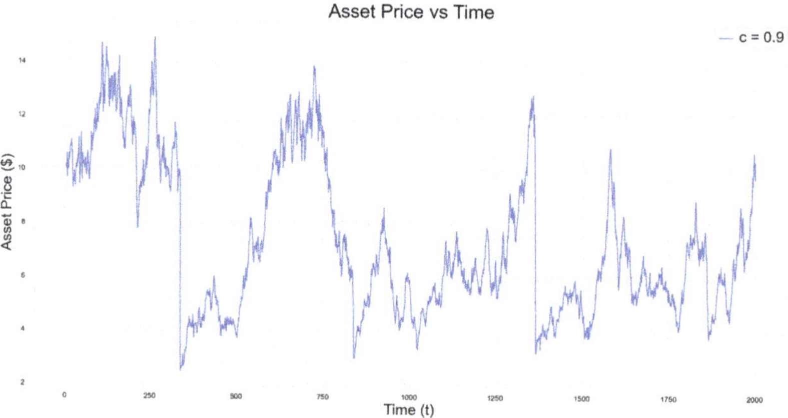

Figure 2. Simulation for $c = 0.9$ for 2000 time-steps. The crash repeats at steps 300, 800, 1300, and 1800.

While in this model financial distress is caused by the presence of excessive transaction costs, Gallegati et al. [78] make it clear that the use of transaction costs is only a modelling tool and many other mechanisms can produce the same result. We consider an alternative scenario with an external field factor, γ, introduced in the utility function, representing the market sentiment and we show that, similar to high transaction costs, it too can cause market shocks more serious than those that might be expected for an equivalent normal distribution. As the number of equity pairs with statistically significant TE values does change over a m

In the presence of market sentiment, the utility function evaluated by the agents becomes

$$U_{it} = (\bar{p}_t - p_{t-1})w_{it} + Jw_{it}w_t^e + G + \epsilon_{it} \tag{8}$$

where $G \sim \mathcal{N}(\gamma, 0.01)$, which means that under normal circumstances ($\gamma = 0$) the utility to buy or to sell is not affected, but under exogenous influence the agents have a higher propensity to either buy or sell. We consider a scenario in which an external event occurs at timestep $t = 300$ and as we can see in Figure 4 this causes a steep decline in market prices, although at a slower rate than the ones caused by financial distress.

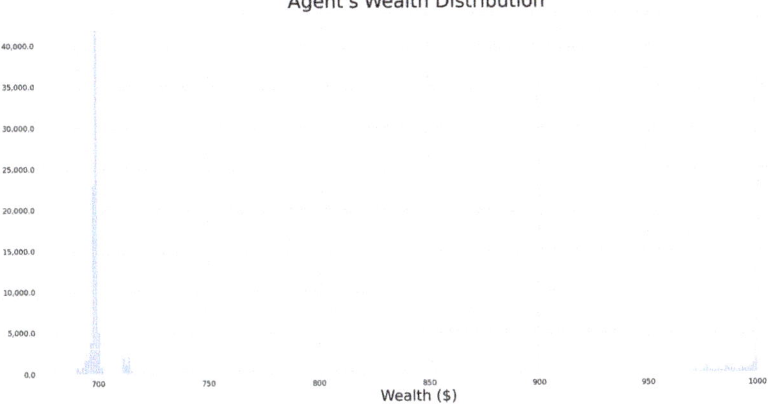

Figure 3. The distribution of wealth over the simulation is concentrated into both the initial point and moment before the crash.

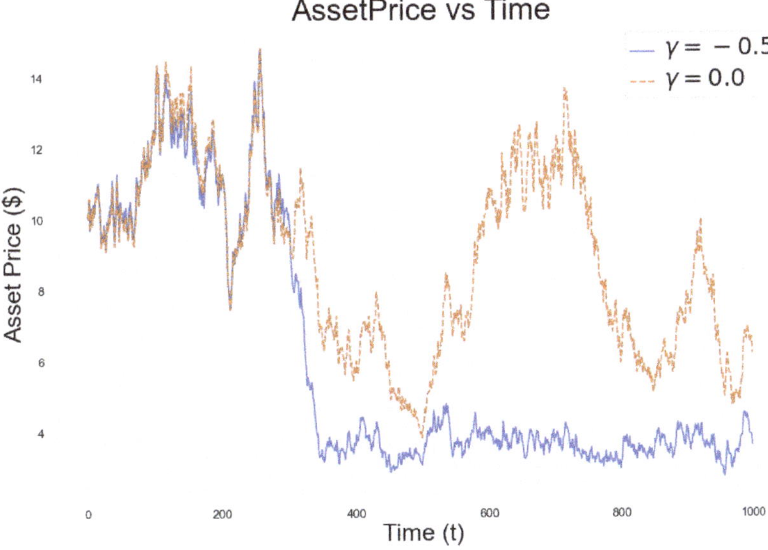

Figure 4. The market sentiment changes to $\gamma = -0.5$ at step 300, causing a depression in the market.

2.3. Remarks

Agent-based models are a powerful tool to simulate the interactions between heterogeneous agents in complex economic environments and to test hypotheses about emergent behaviour originating from those interactions. In this model, Gallegati et al. extended their earlier work by the inclusion of financial constraints at the individual agent level, which endogenously induces a strong nonlinear dynamic. Such simulations are an important tool for calibrating our intuition, in terms of interactions and constraints, for the purposes of policy-making as they can be used to evaluate alternative scenarios and guide decisions that can lead to better policies. However, these are only a guide, and what is needed to extend this work to a more practical understanding of policy-relevant parameters are agent-based models that can also be used to perform more fine-grained scenario sensitivity testing using real data tuned to a specific market as we demonstrate in the next section.

3. Trading Houses: An Agent-Based Analysis of Stressed Markets

Our recent work looking at the Australian housing market [62] has shown that from 2016 onwards this market has been in a volatile state in which high housing prices have been coupled with higher than usual fluctuations in their values, an aspect of the Sydney market that is not unlike the periods of financial distress covered in Section 2. This behaviour contrasts with the previous decade (between 2006 and 2016) in which there were periods of both growth and decline, but the trend has been much less volatile. In this work, we argued that the current state of high uncertainty has been caused by a combination of two factors: the households' trend-following aptitude, i.e., their tendency to 'market herding' behaviour [131,132], and their collective propensity to borrow. The former behaviour is quantified by a parameter that reflects a household's desire to follow the price trend, as accounted by the balance of the monthly costs associated with acquiring a house and the anticipated long-term gains in house value due to market growth. The latter behaviour is quantified by either the mortgage rate or by the observed statistical relationship between a household's income and mortgage. Using a multi-agent model that has realistic data and dynamics that are known to follow real market prices, we were able to probe the model for the mechanisms that drive this new behaviour, without changing the underlying interactions and mechanisms in the model.

It is important to realise that the higher volatility in market prices have been observed not only in the modelled market, but also in the actual market. Such uncertainty is typically an indication of a critical transition, when the system approaches a bifurcation point that separates two (or more) states with relatively stable dynamics such as that studied by Scheffer et al. [133] or the period of financial distress of Gallegati et al. As the system is composed of a large number of interacting agents, taking account of all the factors that influence the evolution of each individual agent—even in the real market—is too difficult a task, resulting in the behaviour of agents being essentially indistinguishable from a stochastic process. This individual stochasticity is compounded by the stochasticity associated with the emergent collective behaviour of a large number of agents. Such phenomena have been observed in other systems [134–136] in which the details of individual interactions between agents are unknown and possibly not even measurable for practical purposes, yet their collective behaviour can still be coherent overall and may provide indicators of the existence of a bifurcation point and the associated systemic risks. Our understanding of such phenomena in socio-economic systems is generally less clear than in these other systems and it is an active area of research.

Agent-based modelling provides a tractable computational tool study stochastic social systems, and housing markets are one of its feasible applications. In these models, we mimic the actions of real households subjected to market conditions (e.g., mortgage rates, housing stock, budgeting constraints, etc.). This allows us to not only identify possible causal relationships between the parameters and the observed market behaviour (e.g., price or population distributions [63,64,137]), but also investigate various alternate realities, i.e., *what if* scenarios, which otherwise are not available for direct experimentation, unlike other applied sciences.

3.1. Simulation Results

There are two elements that are essential specifically for housing markets, affecting the observed price dynamics: (1) the proportion of income expenditure on non-housing consumption and (2) deciding whether or not to buy a house and at what price point.

We investigate the effect of COVID-19-related government policy interventions by exploring alternative financial realities compared to the one people experienced during the 2016–2019 period, which we refer to as the *Baseline* model [62]. In particular, we consider three alternative realities: denoted as the *Rate*, *Income*, and *Liquidity* realities. In the alternative *Rate* reality, the mortgage rate is lower by 2 percentage points compared to the baseline (e.g., from 5.3% to 3.3%). This reflects the reduction of the base rate by the Reserve Bank of Australia and the corresponding reduction of mortgage rates by banks. In the alternative *Income* reality, the proportion of income households pay to non-housing consumption is reduced by a factor of two, compared to the baseline (i.e., from 60% to 30%). This models the effects of a large portion of households being left without an income due to work restrictions. In the alternative *Liquidity* reality, the fraction of accumulated wealth households pay to non-housing consumption is reduced to nil (from 0.25% in the baseline model). This is another aspect of household stress when the population reduces daily spending. Importantly, in these alternative realities, all other parameters of the *Baseline* model are held constant, including households' collective assessment of the market (quantified by the trend-following aptitude) or various house-related taxes.

The results of the simulations are presented in Figure 5. We see that in each of the alternative realities the nature of the price dynamics–high volatility and an upward trend— is similar to the baseline. This is due to the fact that all alternative realities exist within the same fundamental conditions of the 2016–2019 market, namely, high trend-following aptitude and high propensity to borrow (see details in the original paper [62]). Yet, in the *Income* reality, we observe slightly lower price volatility compared to the other realities, which reflects higher certainty in the price trend.

We next focus on the differences between the price trends in the alternative realities and the baseline model in Figure 6 by setting the baseline model's index equal to zero so that the *relative* price trajectories of the other realities is made clear. Here, we see that changing household spending attitudes with respect to the accumulated wealth (as in the *Liquidity* reality), does not affect the price level significantly. In contrast, changing the spending attitude with respect to income (as in the *Income* reality) gradually increases the price level by \$40–60 k (thousand Australian dollars) over the course of the simulation. Furthermore, decreasing the mortgage rate (as in the *Rate* reality) results in increasing the price level by \$50–70 k. With the baseline price level of \$1150 k, the *Rate* and *Income* alternative realities result in an increase of ~5.2% in overall house prices. This result is consistent with the year on year price increase in Sydney from \$1135 k in December 2019 to \$1211 k in December 2020,[4] an increase of ~6.7% where the difference might be explained by the combination of multiple factors during COVID-19.

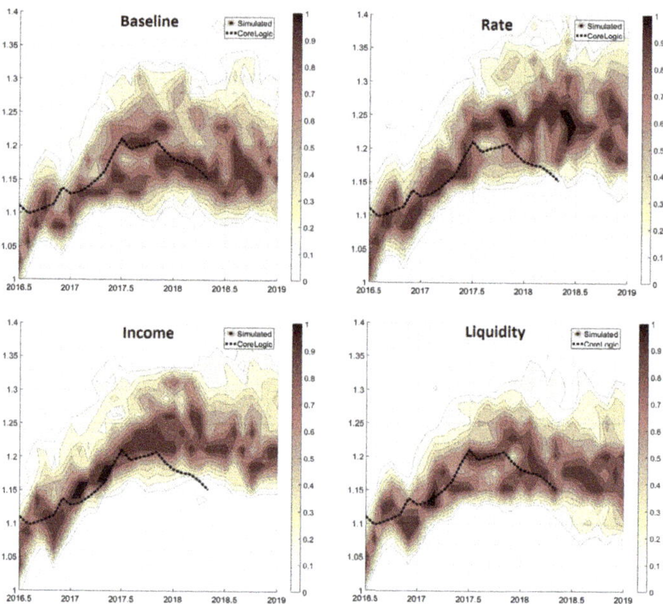

Figure 5. Histogram distribution of the prices from an ensemble of 64 simulations, for the alternative realities and the baseline model. Black line corresponds to the running average of monthly averages of the actual sales price and is the same in each plot, made available from Securities Industry Research Centre of Asia-Pacific (SIRCA) on behalf of CoreLogic, Inc. (Sydney, Australia).

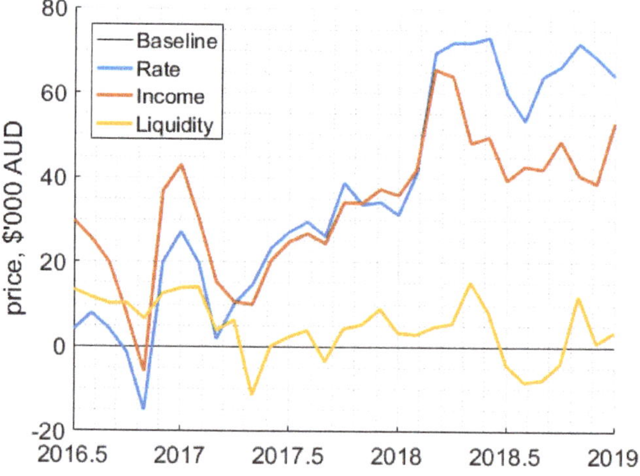

Figure 6. Difference between the median house market prices in three alternative realities and the baseline model.

3.2. Simple versus More Complex Agent-Based Models

We should emphasise that in obtaining these results we did not perform any further calibration of our earlier published model [62]. This tells us that our agent-based model is capable of tracking very fine differences in household constraints that are out-of-sample with respect to the original calibration of the model.

In adjusting the model to reflect alternative economic realities we are changing the constraints placed on the agents using the same principles as described in Section 2 but now these constraints have readily identifiable interpretations in terms of a real market. In particular, constraints such as 'budgetary limits', 'taxes', or 'interest rates', constraints that have been proposed in stylised models [56,78], can be tested, calibrated, and validated against real market data, providing a quantified foundation for informing policy decisions, rather than models that have earlier been more qualitative in their description of market features. It is this step from theoretical and simplified models to richer, data informed, simulations that will make these types of models much more useful in the future.

4. Fluctuations in Equity Markets at Crises Points

An index of a financial market for trading in equities, such as the NASDAQ, the Dow Jones Industrial Average (DJIA), or the Standard and Poors 500 (S&P500), is often taken as a broad indicator of a country's economic health as it can be understood as the 'market's perception' of the economic performance of the industries in which the equities are traded. If the market, as measured by an index, is growing strongly then the underlying businesses are often thought to be growing strongly as well, while a declining index is often taken to be an indicator of poorly performing businesses and consequently a poorly performing economy. For example, the S&P500 is a US-based index of the 500 highest capitalised stocks on the New York Stock Exchange and so it is seen as an indicator of health of the US economy. By following it one can get a feel for the relative performance of the economy over time.

However, individual stocks and their contributions to the overall dynamics of an index have also been studied collectively as indicators of a market in sudden crisis. See, for example, the use of Pearson cross-correlations between equities studied by Onnela et al. [108,109] and the use of mutual information by Harré et al. [55] to study the non-linear dynamics of markets near crises. This is similar in principle to the measurement of neural dynamics during and epileptic seizures [138–140] or, at the aggregate level of entire systems, measuring the statistical signatures of tipping points in ecological and climate time series [141–143]. Other measures have also been used to study market crises and their potential to measure systemic risks in financial markets, for example transfer entropy, a measure of the temporal cause-and-effect of price movements, similar in nature to Granger's causality [113], has been used to study the Asian market crisis of 1998 [70]. A table of the different measures and their relationships is illustrated in Figure 7, and see the references just given for a more detailed description of these methods. This is an active area of research in which new developments are unfolding in multiple areas. In what follows we will use transfer entropy (TE), a measure of Granger-like causality [144], to examine inter-price dynamics in equity markets and we will simply refer to the causal relationship from equity Y to equity X as $\mathbf{T}_{Y \to X}$. For an introduction to its use in general see the book *An Introduction to Transfer Entropy* [71] and for its use in economics see Chapter 6 therein. Here, we introduce, in order of appearance, the entropy, joint entropy, and conditional entropy and then use these to define the transfer entropy:

$$\mathbf{H}(X) \quad = \quad -\sum_{x \in X} p(x) \log(p(x)) \tag{9}$$

$$\mathbf{H}(X,Y) \quad = \quad -\sum_{(x,y) \in (X,Y)} p(x,y) \log(p(x,y)) \tag{10}$$

$$\mathbf{H}(X|Y) \quad = \quad \mathbf{H}(X,Y) - \mathbf{H}(X) \tag{11}$$

Now, we can define the Transfer Entropy between two time series $\{X_t\}$ and $\{Y_t\}$ as the difference between the entropy of $\{X_t\}$ conditioned on its lag-1 history $\{X_{t-1}\}$ minus the entropy of $\{X_t\}$ conditioned on both the lag-1 history $\{X_{t-1}\}$ and $\{Y_{t-1}\}$:

$$\mathbf{T}_{Y \to X} \quad = \quad \mathbf{H}(X_t|X_{t-1})) - \mathbf{H}(X_t|X_{t-1}, Y_{t-1})) \tag{12}$$

sometimes the argument to the entropies are made clearer by explicitly stating the distributions as we do below and see [71] for generalisations to different lags and further conditional factors for the probability distributions. In plain language, the TE measures the amount of information that flows from Y_{t-1} to X_t once the history X_{t-1} is accounted for. We implement this method for financial time series in the next section.

4.1. Analysis Using Transfer Entropy for the DJIA Market Shocks

The DJIA is a price-weighted index of equities for thirty of the most prominent industrial firms in the United States and as such is seen as a very broad measure of the manufacturing health of the US economy. Here, we look at this index of companies at periods covering three market events: the Federal bailout decision of 2008, the flash crash of 2010 and the COVID-19 crisis of 2020 together with three control dates where there are no recent events of any significance. For each of these market events, the analysis focuses on the day before the event, the day of the event, and the day after the event with three thirty-minute periods (backwards from 11:00, 13:30, and 15:30).

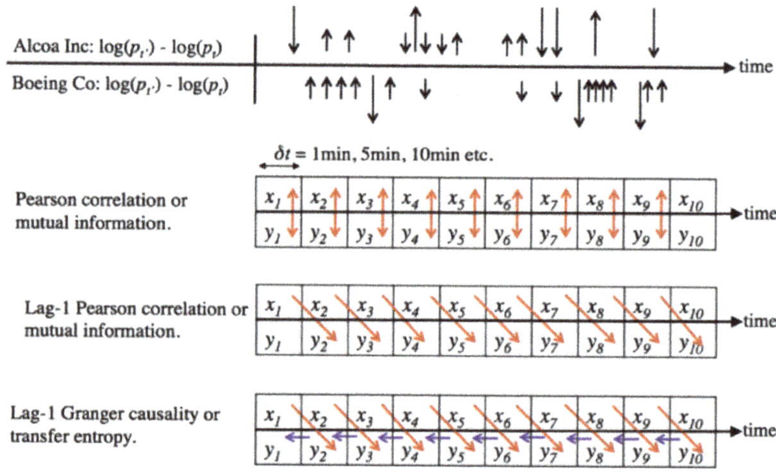

Figure 7. Top line: the log differences in price movements is where the structural analysis of market movements starts, here Alcoa (x in the lower diagrams) and Boeing (y in the lower diagrams). Subsequent lines: Multiple methods have been used to calculate the co-movement relationships between equities using price fluctuations. Originally discussed in [70].

Across each of these periods, trade data for each stock were aggregated into one minute averages: d_t^i so that within a 1 min interval $[t, t+1]$ we simply calculate the average price within that interval. The result is a time averaged binning of the continuous trade data over 1 min intervals and we denote this: $\{d_t^i\} = \{d_1^i, d_2^i, \ldots, d_T^i\}$. Subsequently, these one minute intervals are used to calculate the TE over three 10 min periods within each thirty minute window. Figure 7 illustrates the method and compares it to other common methods, but in the notation we have here for two time series of equities $\{d_t^i\}$ and $\{d_t^j\}$ the TE is

$$\mathbf{T}_{j\to i} = \mathbf{H}(p(d_t^i)|p(d_{t-1}^i)) - \mathbf{H}(p(d_t^i)|p(d_{t-1}^i), p(d_{t-1}^j)) \tag{13}$$

Note that this form is extended using the KSG (Kraskov, Stögbauer, and Grassberger) algorithm for effectively estimating probability distributions. A full treatment of this approach is available in *An Introduction to Transfer Entropy* [71].

Finally, the statistical significance is evaluated by taking the trades within each interval and reshuffling them 100 times to establish a surrogate test of the Transfer Entropy with the temporal relationships randomised. This is used to estimate the probability that the statistical significance of the obtained Transfer Entropy by reference to the randomised samples using a 0.05 p-value test for significance.

The following heat maps show the equity pairs with statistically significant values of TE, as measured by the software package JIDT [145]. Figure 8 shows the result of measuring the pairwise TE between equities in the DJIA. In each of the four time periods (each having a 3×3 matrix of heat-maps, top-left, top-right, bottom-left, bottom right) there are three days stacked from top to bottom (the day before the event, the day of the event, and the day after the event) and during each of these days there are three time intervals (11:00, 13:30, 15:30). Each of the 36 heat-maps is a matrix of TE from each of the 30 equities in the DJIA (rows) to each of the equities in the DJIA (columns). Periods with an increased occurrence of TE are highlighted with a light blue frame. The colour of each pixel is an indication of the size of the TE value. Periods with high maximum entropy are highlighted in red.

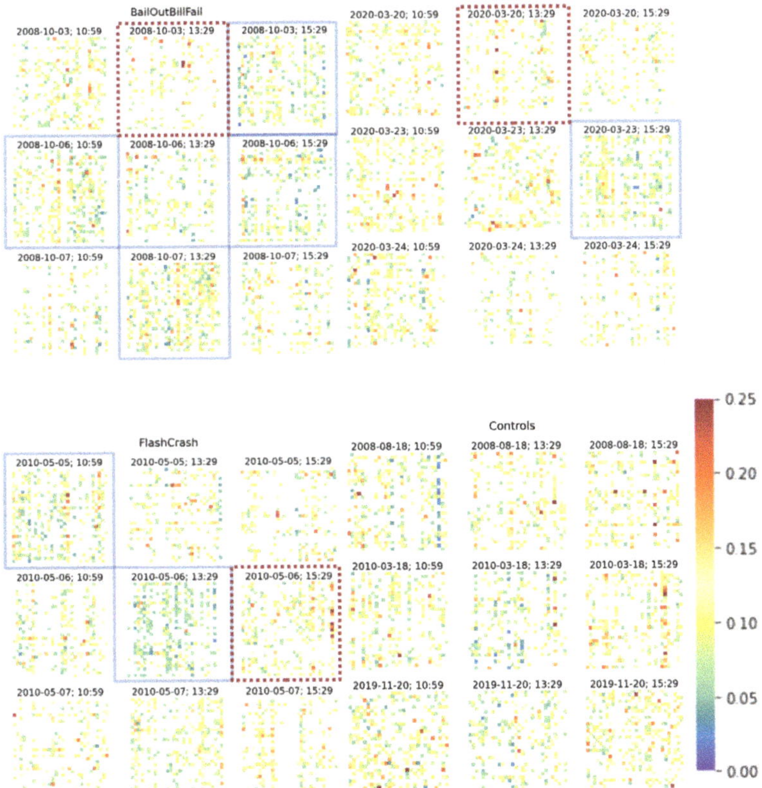

Figure 8. Heat maps of TE for three 30 min periods on three days around a market event. From top left to bottom right: The Federal bailout package failure during the Global Financial Crisis, the COVID-19 crash of 2020, the flash crash on the 5 May 2010, and three control dates on which nothing happened.

Looking at the event periods in Figure 8 qualitatively, entropy transmission is comparatively lower the day prior to each market crisis for all three cases. However, within those periods of fewer instances of statistically significant TE, some individual pairs produce very high values of TE. For the COVID-19 crisis, the total count of statistically significant values of TE is highest in the late afternoon of the day of the crisis (the 'event') although the size of the transmissions shows the largest spikes in transmission occur in the morning and early afternoon intervals. Considering the results in Figure 8 at the three market events, there are increased incidences of equity pairs with statistically significant levels of TE, however there are lower values of 'peak' TE leading to a 'blue hue' for those periods of heightened activity with more active pairs but fewer high values of TE.

Looking at the count of the number of equities with high TE values during a market event in Figure 9 shows behaviour significantly different from the control sample. In the period at the end of the day prior to each market event, the occurrence of high values of TE is reduced and remains lower overall than the control periods. However, within that prior afternoon before each of the market events, there is a peak of activity when statistically significant TE values occur more frequently. This can be observed by looking at the maximum transfer entropy value for a period across the market events in comparison to the control data, seen in Figure 9.

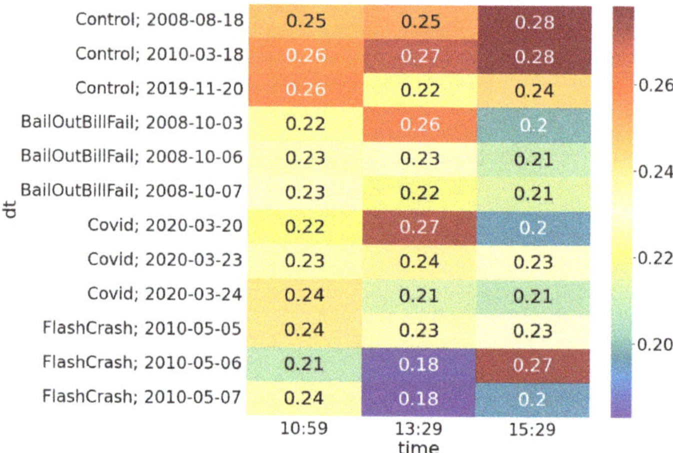

Figure 9. Heat map of the of indices with maximum transfer entropy and a p-value < 0.05 for all dates.

Looking at the distribution of TE events for equities in these periods in Figure 10, the profile is similar. Across the day, an estimate of the distribution of counts of statistically significant TE values (using a Gaussian kernel density estimator) gives a peak count of transfer entropy events at any one time of twelve with a transfer entropy level of 8 to 9 bits.

However, analysis of individual periods within and around the market events in Figure 11 provides additional characteristics which distinguish the market events from the control windows. In particular, for the flash crash of 2010 we see a peak number of equity pairs with entropy transfer as an event occurs but that peak is skewed to the left indicating a decrease in the number of equity pairs with higher TE levels.

For each market event there is at least one period where the distribution has an increase in central tendency. Kurtosis measures used to capture this property have been found to be useful in these cases.

Note that this increase may not always result in changes in the kurtosis. This is because kurtosis captures the extent to which the tails of a distribution contain values greater or lesser than those that might be expected for an equivalent normal distribution. As the number of equity pairs with statistically significant TE values does change over a market event, an increase in the central tendency for the distribution is possible without a change to the tails of the distribution and the attendant change in kurtosis, see Figure 12. However, a reduction in kurtosis was observed for the population of equities undergoing transfer entropy relative to the control data (Figure 13).

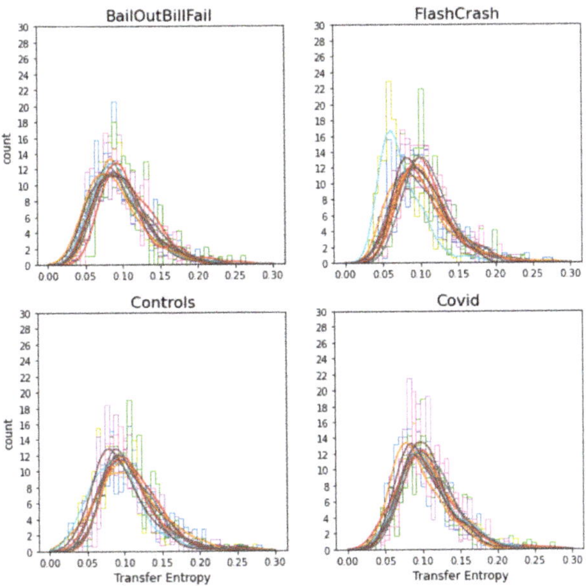

Figure 10. Transfer entropy distributions for market events. Distribution of transfer entropy for three market events and a control sample.

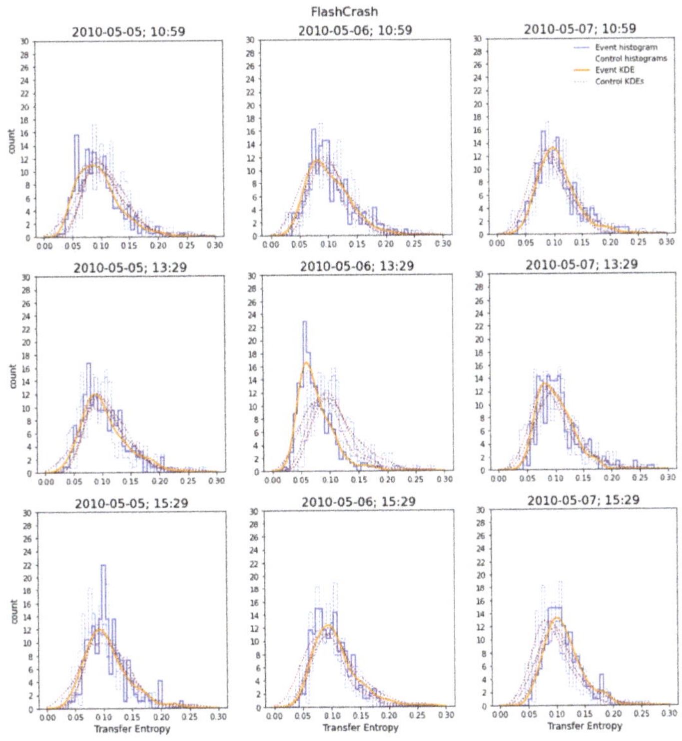

Figure 11. Transfer entropy distributions for market events. Distribution of transfer entropy for the Flash Crash market event over three 30 min intervals over a three-day period.

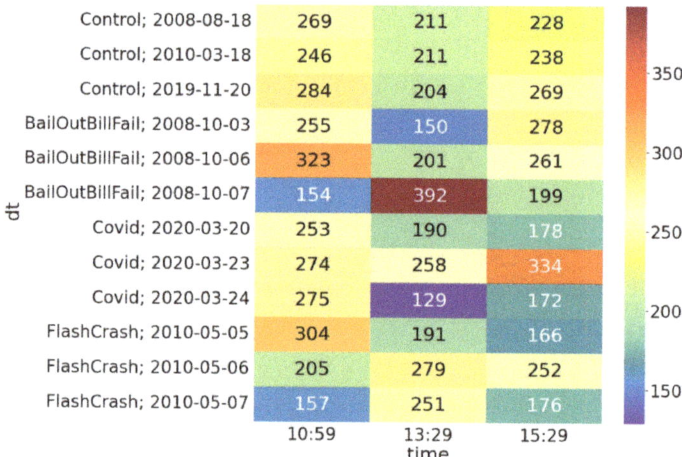

Figure 12. Heat map of the count of equities with transfer entropy with a *p*-value < 0.05 across key dates.

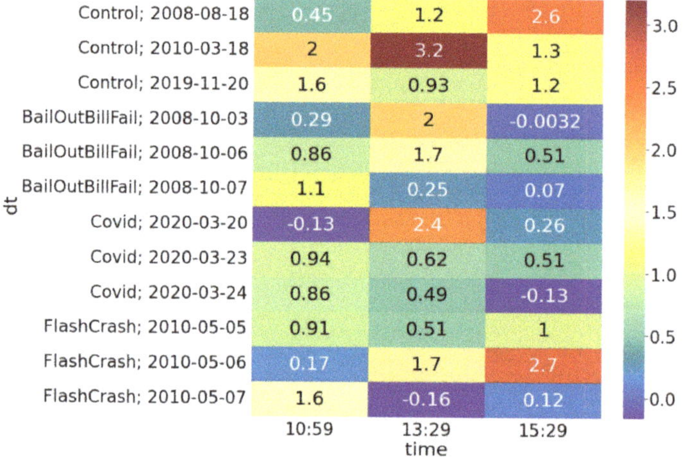

Figure 13. Heat map of the Fisher kurtosis from the distribution of equities with transfer entropy with a *p*-value < 0.05 across key dates.

4.2. Remarks

There is evidence of features in the transfer entropy activity in measures taken in and around the market events reviewed. There are similarities that can be shown with TE measures for the GFC, the May 2010 flash crash, and more recently the COVID-19 crisis. The most fruitful insights have been on an aggregate level rather than looking at calculations for individual stocks. This is in line with earlier results for the Asian financial crisis that used the same methods [70], and more generally we might expect changes in the statistics of time-series near or at a crisis point as shown in other systems [133,143] due to their very nature being that of a non-stationary event, so although these earlier studies use univariate time-series it is an interesting observation to see these changes at the multivariate and interaction level of analysis. It has also been possible to demonstrate real difference between a control population and the data from significant market events, immediately before, during, and after those events. The challenge now is to refine the calculations, gleaning clearer results with new parameters, a move from more qualitative observations to quantitative analysis and the search for real predictive capability allowing the understanding of market events through the prism of transfer entropy and associated ideas.

5. National and International Trade in Value Added

Analysing international trade as a complex network of interactions provides useful insights into the structure (topology) and dynamics of world trade and has been used to map out recent changes in trade relations. In an article by Fagiolo et al. [146], network statistics were used to determine the importance of links within the weighted global trade network between 1981 and 2000, and it was found that the majority of links were relatively weak although countries with more intense trade relationships are more clustered together. In another study by Bhattacharya et al. [40], it was shown that over a 53-year period to the year 2000 the 'rich club' controlling approximately half of the world's trade has been shrinking. In a similar vein, Maeng et al. [147] used minimum spanning trees (MSTs[5]) to show that international trade networks were dominated by strong links between hubs of larger economies such as the USA, Germany, and China. Further progress was made by Barzel and Barabási [148] when they developed a theoretical framework (independent of its application) that uncovered universal properties of the relationship between network topology and network dynamics. This was the first self-consistent theory of dynamical perturbations in complex systems that could systematically separate out distinct contributions from the topology and the dynamics. Andrea Aria [149] from the European Central Bank has also explored how during the Global Financial Crisis the elasticity of goods exports was vastly different to that of services exports. Within such a large and varied range of new results there is considerable scope for new developments. In what follows we extend some of these ideas to networks of 'Trade-in-Value-Added', a relatively uncommon measure of traded value between industries and countries, in order to extract key qualitative features. For more concrete policy implications of network analysis see for example the work being carried out at INET [150].

5.1. Value-Added Trade Networks

In what follows, we look at the network topology of country and industry-based trade using national and international Trade-in-Value-Added (TiVA) tables from 2005 and 2015 made available by the OECD. Nodes in a network can be either countries or industry sectors, links are between nodes are directed as they can be either originating from or terminating on a node, and in general they are not symmetric. From this we can formally represent a network as an asymmetric square matrix where the matrix entry for row i and column j, $M_{i,j}$, represents a transfer of value from node i to node j, where generally $M_{i,j} \neq M_{j,i}$.

The most common form of trade network analysis is based on *Gross Trade* (see the left hand diagram in Figure 14). In these networks, the cost of each good or service, regardless of whether or not it is a component in the production of another good or service (i.e., an intermediate input), is its cost of purchase which implicitly includes the cost of the intermediate goods and services used in its production. These value chains describe cumulative flows through trade networks but they offer no insights into how much value intermediate goods and services provide to the final goods and services purchased by consumers, also called final demand. Final demand plays an important part in the analysis of the national impact of an economic shock because it is used to calculate a country's Gross Domestic Product (GDP), something that cannot be done directly from gross trade data.

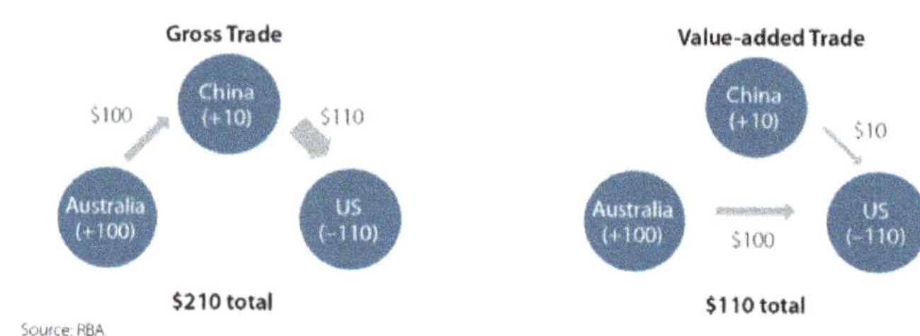

Figure 14. A comparison of gross trade with trade in value-added between countries, a similar diagram holds for particular market segments as well. Example from the Reserve Bank of Australia report [151].

To help address this, the OECD reports an alternative measure, the Trade-in-Value-Added [152], that records how much intermediate value is provided to final (i.e., consumer) demand for a good or service (see the right hand diagram in Figure 14) by both industry and country, see Figure 15. The point to note is that summations of links in gross trade networks will not reflect the true contribution of each market sector or country to GDP (compare the two totals shown in Figure 14: $210 vs. $110). Furthermore, in the analysis of economic shocks we need to distinguish between intermediate value production, which is related to employment and supply, and consumer consumption, which is related to demand, as each industry sector is made up of both intermediate products and final demand for products, and these are not symmetrical relationships but they are jointly captured in the TiVA tables. This is central to understanding the interconnected consequences of an uneven supply and demand shock like COVID-19. For example, in the article by del Rio-Chanona et al. [150], they were able to estimate

1. supply-side reductions due to the closure of non-essential industries (which can be captured in part by the intermediate value added in TiVA tables), and
2. demand-side changes caused by individuals immediate response to the pandemic, such as reduced demand for goods or services that are likely to place people at risk of infection (which is captured by final demand in TiVA tables).

		Intermediate Consumption			Final Demand			G.O.
		Country 1	...	Country N	Country 1	...	Country N	
		Ind. 1 ... Ind. K	...	Ind. 1 ... Ind. K	FD ' ... FD F	...	FD ' ... FD F	
Country 1	Ind. 1 ⋮ Ind. K	Z^{11}	...	Z^{1N}	Y^{11}	...	Y^{1N}	X^1
...	⋮	⋮	⋱	⋮	⋮	⋱	⋮	⋮
Country N	Ind. 1 ⋮ Ind. K	Z^{N1}	...	Z^{NN}	Y^{N1}	...	Y^{NN}	X^N
Value Added¹		W^1	...	W^N	Taxes less subsidies on final products			
Gross Output		X^1	...	X^N				

Figure 15. The original OECD TiVA table structure available from their website[6]. The final demand data in green is used to construct the networks below.

From the matrix shown in Figure 15, we can see that the OECD tables can be split into two parts: the internal economic structure of a country, represented by the trade between industry sectors within each country, and the global trade in value, which we will capture by aggregating total values traded between countries. In so doing, we can begin to understand how both local and global value chains are impacted by economic crises.

5.2. Features of Australia's Internal Trade Patterns

Looking at the local value chains for Australia, we study the total contribution each market sector makes to every other market sector by summing the entire row of each industry (in the TiVA data tables) and in measuring the contribution made to an industry we sum each industry's entire column (also see Figure 16 below). This captures the total consumption of value (final demand) of each market sector and the total production of value (value added) of each market sector, and so we can plot the relationship between these two aspects of a market sector by value on a single diagram, as shown in Figure 17.

Figure 16. Australia's Domestic Industry Sensitivity Matrix, white cells represent low value entries and brighter green entries represent higher value entries. Diagonals have been set to 1.

In Figure 17, we analyse Australia and compare it with that of the U.S. using data that were obtained from the OECD's TiVA database of input–output tables.[7] In these plots, the x-coordinate is the total value of final demand of value in each market sector: it is the sum of all value contributed from other market sectors that contribute to the final demand of each individual market sector indicated on the plot. The y-coordinate is a proxy for the supply side of value each market sector provides to the final demand of the rest of the economy, calculated by summing the value added by each sector to all other sectors.

In Australia, we observe that the supply and demand is reasonably well balanced: sectors with low values of total inputs are closely related to sectors with a low value of total outputs to the rest of the economy. On the other hand, high-valued input sectors also have high-valued outputs ($R^2 = 0.6626$). In these terms, supply and demand of value in the national value chain of production are reasonably well balanced and a shock to each sector is likely to have an equivalent impact on both the supply side and the demand side of value. The two notable exceptions are education and defence that have a relatively high supply side of value but a relatively low value of demand side of value.

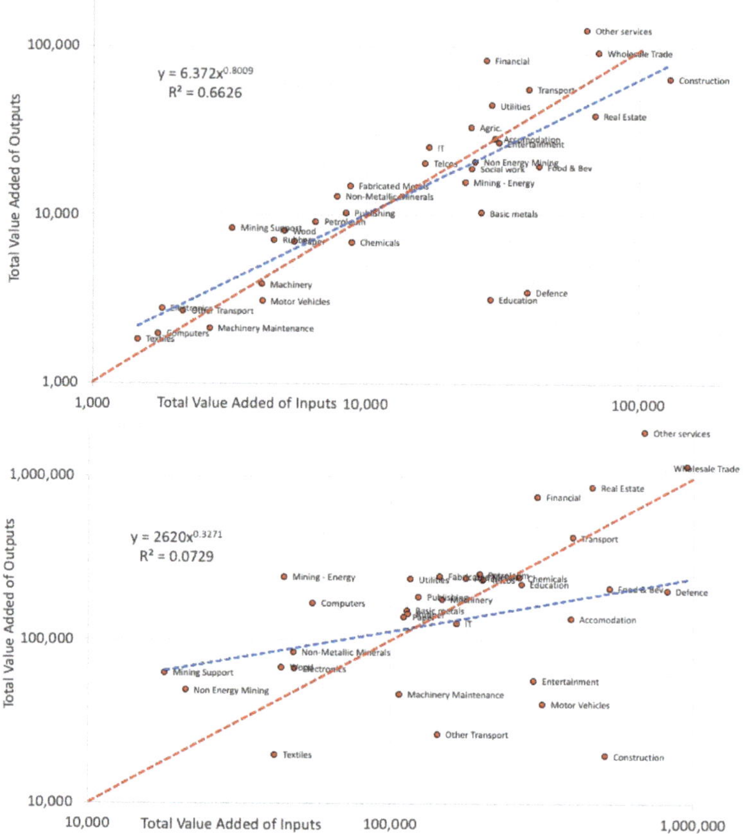

Figure 17. Log-Log plots of industry input versus industry output in millions of U.S. dollars for trade in value added. Top: "from each industry" versus "to each industry" for sector within the Australian economy. Bottom: "from each industry" versus "to each industry" sector within the U.S. economy for comparison. The blue-dashed lines are the Log-Log regression (with equations and R^2 values shown) and the red-dashed line is inserted by hand with a gradient = 1.

The U.S. tells a very different story to Australia in terms of its internal economic structure. There is very little relationship between the demand side of value and the supply side of value ($R^2 = 0.0729$), and a number of industry sectors such as entertainment, construction (and other heavy industries), and transport have greater demand of value than supply of value. On the other hand, real estate, wholesale services, and other service are more similar between the two economies (once differences in the total size is accounted for). Of course this is only the internal structure of the economies and the relationship between industry sectors and their international trading partners, as expressed in the value of overseas final demand and overseas contributions to internal final demand still need to be accounted for in order to have a more complete economic picture.

5.3. Predicting the Impacts of Exogenous Shocks

Now that we have a framework for understanding the structure of trade, we want to analyse how this structure might be impacted by exogenous shocks such as COVID-19. The dataset is reduced to Australian-only input and output industries using a *Domestic Industry Sensitivity Matrix*, see Figure 16. The matrix is a topological representation of the trade input-output function of each industry which allows quantitative, structural analysis of important relationships between industries as well as the impacts of first and second-order flow-on effects from perturbations in inputs or demand functions. For example, we can see that Transport, Utilities, and Wholesale Trade contribute value to final demand for almost every other sector in Australia, and Construction, Defence, and Education receive value to their final demand from a large part of the Australian economy.

To demonstrate how the sensitivity matrix can be interpreted, we analyse the first-order demand and supply effects of a 10% reduction in construction output. Tables 1 and 2 summarise the top 3 most-impacted industries from a notional value basis and a relative basis (as a concentration of impact that industry has relative to trade with all other industries).

As the fall in construction affects the demand for input industries, we can analyse the subsequent output effects of a fall in construction on industries that rely on construction as an input.

These are just the first-order supply and demand effects of a temporary perturbation in the demand and supply for an industry based on the internal trade of the Australian economy. Longer-term impacts will propagate into second-order effects which can be analysed using the sensitivity matrix.

Table 1. Summary of demand-side impacts on total industry input.

Demand Impacts			
Industry Affected	**Notional Value Affected**	**Industry Affected**	**Effect as % of Industry Total**
Other services	−2850	Non-Metallic Minerals	−9%
Wholesale Trade	−1667	Wood	−8%
Non-Metallic Minerals	−1107	Fabricated Metals	−6%

Table 2. Summary of supply-side impacts on input industry total.

Supply Impacts			
Industry Affected	**Notional Value Affected**	**Industry Affected**	**Effect as % of Industry Total**
Real Estate	−2337	Real Estate	−3%
Other services	−533	Wood	−2%
Defence	−514	Utilities	−1%

5.4. Features of Global Trade Patterns

The global patterns of trade between industry sectors and countries is a complex, multiply layered network of interactions and so to simplify our analysis we only study the total values of trade between countries. In Figure 18, we show the total trade in value added between countries for 2005 (left) and 2015 (right) from which we can extract some qualitative features, noting that the node representing the rest of the world (ROW) has been held approximately constant so that we can compare relative changes in the other countries.[8] We can clearly see that the USA's export influence has declined in these 11 years while China has increased, matching the results of other work, for example, Deguchi et al. [153] have reported that between 1992 to 2012 the USA decreased in global trade authority while China has increased and recently surpassed the USA in this respect. We can also see that the ROW has become a more central element in the network, having a weaker relationship with the USA but a stronger role to play in trade with other region of the world. Other features are also evident: Japan and a number of European countries have decreased in value added trade while Korea has increased. One aspect of the structural change in trade dynamics has been the emergence of intraregional trade and regional supply networks, see, for example, Kelly and La Cava [151] and Zhu et al. [154]. This can be seen in Figure 18 where the global regions are colour coded and segmentation of the networks by regions is apparent.

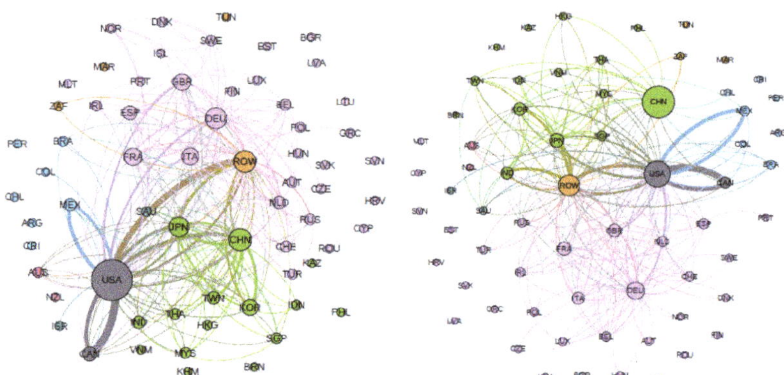

Figure 18. World trade network of value add. (**Left**) 2005 inputs to countries for final demand from intermediate products. (**Right**) 2015 inputs to countries for final demand from intermediate products. Country nodes are sized to represent relative differences in total exported value with the rest of the world node (ROW) held approximately constant

5.5. Remarks

The motivation for this analysis has been to explore how a complex network-based system can represent the characteristics of an interconnected global trade network at the global, country, and industry levels. By probing network structures at these levels through matrix sensitivity analysis and changing patterns in global trade, we have a method to explore the potential demand and supply shocks that propagate through national and global networks. Further work can extend these methods to incorporate occupations, geographies, and work activities involved within each industry which would assist government policymakers in crisis response, a process that has already begun with the work being carried out at places such as INET.[9] Another consideration not explored in this research is that of the elasticity of industry supply and demand functions which would impact the magnitude of exogenous shocks, see, for example, the work of Escaith et al. [155] on the impact of the global financial crisis on trade networks.

6. Project Economics and the Knock-On Macro-Effects of Their Delay, Cancellation, or Failure

Central to the economic development of a country is the delivery of innovative products as well as the infrastructure necessary for economic expansion, such as roads and power stations. From the view of an organisation though, projects are an important organisational construct used to plan and control the delivery of a vast array of products and services [124,156–158]. By definition, a project is a temporary activity undertaken to create a unique product, service or result [159]. As temporary forms of organising, they have the potential to generate innovative capacity and strategic flexibility [160]. Quantifying the number and scale of projects under management within a country or across the world is difficult, but based on a study of three Western European countries it has been shown that the degree of projectification of an economy relative to the gross domestic product is in the order of 30% [161]. This provides us with the motivation to better understand the effects of crises on project development and deployment in the context of the economic growth of an economy.

Organisations vary in the degree to which they engage in and describe their work as project focused. Yet, projects are increasingly being employed as a tool of strategic innovation in all industries: for the delivery or development of their products or services for their customers, the transformation of their own structure or culture, or the design and implementation of their strategies [124]. Broadly speaking, organisations can be separated into at least two categories: project-based organisations and project-oriented organisations. Furthermore, some firms offer complex and individualised solutions to their customers that are contracted before project development starts [124]. Turner and Keegan [162] have argued that these firms are project-based because of the customised demand of their clients. On the other hand, some organisations choose to become project-oriented as a matter of strategic choice [124].

Although there is no formal convention differentiating the types of projects undertaken by these organisational typologies, there is arguably an intuitive distinction to be made. If an organisation is project-oriented, the undertaken projects are often numerous and small in scale (e.g., local, short-term, and lower cost). Project-based organisations, in comparison, are often found to be involved in the delivery of large-scale investment projects with a significant degree of complexity. If a project is sufficiently costly (>US$ 1 billion), spans over several years, and is expected to have a significant societal or economic impact, it is termed a megaproject.

Project-oriented organisations may have tens to hundreds (or even thousands) of projects underway for many different clients, internal and external, at any given time. This large number of concurrent projects presents several portfolio management challenges, including the identification of, and intervention on projects that go off track during delivery [60]. In these portfolios, each project, and its expected outcomes, is important enough in its own right to demand the time and resources needed to sustain it, but it is unlikely that the delay, cancellation or failure of one or even a few projects will threaten the overall success or survival of that firm.

The same is not true for project-based organisations, which may have just one or two very large, long, or complex megaprojects under management at any given time. Many other partner, supplier, and subcontractor organisations may be engaged in these megaprojects. Therefore, the delay, cancellation, or failure of any one of these projects may have a significant impact on the survival of the delivering organisations, and have a much broader impact on the economies, environment, or even societies for whom they were being delivered [163]. Flyvbjerg ([125], p. 6) defines megaprojects as being "large-scale, complex ventures that typically cost US$1 billion or more, take many years to develop and build, involve multiple public and private stakeholders, are transformational, and impact millions of people". Megaprojects are used as the preferred delivery model for infrastructure, water and energy, mining, enterprise systems, mergers and acquisitions, space exploration, the development of new aircraft, airports, drug development, national broadband, and Olympic Games [125–127].

6.1. Mega-Projects and the Economy

Megaprojects facilitate the implementation of technological and organisational innovations at a scale that is usually inaccessible to most organisations. Edward Merrow in Flyvbjerg et al. [164] (p. 4) wrote:

> ... such large sums of money ride on the success of megaprojects that company balance sheets and even government balance-of-payments accounts can be affected for years by the outcomes. The success of these projects is so important to their sponsors that firms and even governments can collapse when they fail.

This extreme scale is reflective of the functional complexity of megaprojects which are themselves often initiated to facilitate the productive efficiency and delivery of many other goods and services. It could be argued that megaprojects emerged as a managerial concept to solve the problem of delivering such complex projects where previously existing project management practices were found to be inadequate. As their inception, much has gone into developing and disseminating megaproject theory in a cycle that has neo-Shumpeterian attributes [165,166].

Megaprojects stand out for another important reason: they are plagued with overly optimistic estimates of time, costs, and expected benefits [126,164]. While optimism bias is not unique to megaprojects, given the scale and import of these projects these underestimations have greater impact. This is well illustrated by the cities competing to hold the Olympic Games, which have consistently underestimated and yet these errors have been repeated every four years [126]. In their working paper, Flyvbjerg and Stewart [127] studied the cost overruns of the Olympic Games from 1960 to 2012. They found that the Games projects overran with 100 percent consistency. They explained that other types of megaprojects experience cost overruns from time to time, but none were found to be this consistent. Additionally, Flyvbjerg and Stewart [127] reported that the Games cost overruns of well over 100 percent, was significantly larger than for other types of megaprojects including infrastructure, construction, information, and communications technology.

Further, the environmental and social effects of megaprojects are commonly found to have been miscalculated or not taken into account at all, and surface during construction and operation, potentially destabilising habitats, communities and the projects themselves [164]. Adding to systematic underestimation, decision making around megaprojects is further impacted by deception and delusion, in the form of strategic misrepresentation by project promoters [164], and exacerbated by misplaced political incentives [126] or political ambitions [167]. Referring to infrastructure megaprojects, one of the arguments commonly made by project promoters to commit public funds is that these projects will generate economic growth in a particular region, country or local area, but these expected regional benefits repeatedly turn out to be unquantifiable, insignificant or even negative [164].

When faced with an economic downturn, project-based and project-oriented organisations may be impacted in quite different ways depending on the nature of their projects and their respective products, services, and customers. In their favour, Aritua et al. [168] argue that projects are complex adaptive systems, and as such project managers and their project teams are always reacting to the changing environment around them. On the other hand, some changes to a project's environment are so large and disruptive that this more organic response may not be able to respond adequately to protect the project and its deliverables.

6.2. COVID-19 at the Project Level

Due to their extreme scale and impact, megaprojects may play a significant role in shaping the (socio-) economic processes of an economy. However, the reverse may also be argued, i.e., where there is an economic disruption like that caused by the COVID-19 pandemic, this can precipitate the cancellation or pausing of megaprojects as their funding bodies reassign previously budgeted capital to other more pressing needs with the resultant loss or delay of innovation and value. Early investigations into the impact of COVID-19 on construction projects in the USA [169,170], UK [171], and New Zealand [172] show that within months of the declaration of the global pandemic, construction companies were seeing large numbers of projects cancelled or put on hold. A report from July 2020 shows that in the United States roads and transportation projects alone, projects to the value of US$9.6 billion had already been delayed or cancelled. As of July 2021, the US state of California was reporting 35 cancelled construction projects (US$ 131 million), 580 delayed projects (US$ 6.03 billion), and a further 224 (US$ 6.7 billion) that have been put on hold due to COVID-19 [173]. Between September 2019 and 2020, construction sector employment in the US decreased by 275,000 year on year [174], recovering from peak employment losses between March and April 2020 of approximately one million workers [175].

This disruption to the project delivery pipeline can be expected to cause several levels of disruption. At the local, regional, and national levels, there is the short-term disruption to local spending on raw materials, goods and services to support projects, and the employment of local labour. In the longer term, the discontinuance of these projects may have far reaching impact due to the delay or non-delivery of the expected benefits of the project. This impact could be felt in slowed or limited urban or rural development, agricultural growth or regional tourism to name a few. In this way, the shorter term savings accrued through the cancellation or delay of a large infrastructure project may be significantly outweighed by the longer term losses mentioned here. Further, there is a significant disruption to the delivering organisations, where firms that design, plan, and deliver these projects are faced with a severe disruption or cessation of expected cash flow, discontinued access to the work site and to the labour needed to complete the work. At the same time, if the project is expected to be continued in the future, these firms are faced with several dilemmas on how to continue to make their project financing payments and leases on critical equipment, and how to retain access to skilled labour when it is not clear when they will be needed again.

The broad and almost simultaneous geographic impact of the pandemic has created a unique situation where specialised firms that might ordinarily have moved resources from one project to another when one was cancelled or delayed may now find themselves with very few continuing projects to work with. Faced with this project pipeline contraction, delivering organisations may decide to take protective actions by laying off employees and other cost reduction tactics. These tactics may serve to preserve the organisation in the short term, however, when project work eventually recovers they may find it difficult to recover lost productivity and talent that these temporary reductions caused. The reduction of the work force itself also has an impact on household incomes for the affected workers, thereby affecting the national economy's final demand just as in the case of many other industries, see Section 5 for a more detailed analysis of the impact that job losses can have on an economy.

As some countries begin to emerge from the grip of the pandemic, there are moves to spur economic recovery by initiating large infrastructure projects.[10] Just as the almost simultaneous contraction of project demand caused delivering organisations to make changes in staffing and spending en masse, starting (or restarting) significant numbers of very large projects over a short period will likely stress delivering organisations as they seek to rapidly revive their projects and recover their work forces while simultaneously competing for newly announced projects in these infrastructure spending measures. Further, this restart may cause several other issues, such as the competition for skilled labour resources who would ordinarily move from project to project as they asynchronously started and ended may now be sought by many delivering organisations who are restarting projects at a similar time.

6.3. Remarks

In this section, we have described how the impact of one or more projects being delayed, cancelled, or failing in organisations that deliver multiple simultaneous projects to many different customers were comparatively small as they only make up a small part of a much larger portfolio. In this case, when one project fails, losses are somewhat easier to accept, and now-surplus human resources may be assigned to other projects in the portfolio. However, if a large portion of those smaller projects were being delivered to one of the more heavily impacted sectors, a large portion of the portfolio might need to be suspended, and may not be reinvigorated when the crisis has passed. In this case, the an organisation may not be able to afford to support the workforce during this period of reduced activity, which may result in layoffs and other cost reduction tactics.

The scene is different for megaprojects whose delivering organisations might have just one or a few megaprojects underway. These megaprojects are already risky prospects, and due to their long planning horizons and delivery timelines, they are particularly exposed to extreme events with large negative outcomes i.e., "black swan" events [125] such as COVID-19 or the 2008 Global Financial Crisis. Despite their devastating effect when they occur, megaproject (and indeed smaller project) managers generally ignore the possibility of black swan events in their planning [125]. We discussed above how big a megaproject might be, and how reliant the delivering organisations and even governments can be on a project's success. Given the economic impact on the region as well as the local and global workforce, it may not be possible to put a megaproject on hold. However, even if the project has the funds to proceed, it may be faced with other issues due to social distancing requirements that affect work site staffing, or the lack of access to expert resources who may not be able to travel to the site of a project, or shortages of raw materials as global supply chains respond to rapidly increasing demand, stressing already compromised global value chains (which we discuss further in Section 5). Even if a megaproject could be stopped altogether (which is unlikely to be a contractual option [125]), in times of economic downturn the injection of capital into the economy through local employment and other locally procured services might provide some stability to augment national level stimulus packages.

7. Conclusions

While all economic crises are idiosyncratic in the details of their cause and effect, the interactions between heterogeneous agents are integral to understanding the life cycle of a crisis-instigated market failure. This point is central in both the general development of 'complex systems theory', see, for example, Barzel and Barabási's work [148] on the interplay between topology and dynamics for perturbations to networks, as well as specifically in complexity economics, for example, Arthur's recent overview [21]. There are two ancillary arguments captured by the present article in support of this primary claim. First, the particular way in which agent heterogeneity is expressed is crucial in the precipitation of market conditions. This is because the topological structure of economic relationships is significantly influenced by the type and degree of difference displayed by agents. Second, in the absence of complete information, market interactions with an inter-temporal component are significant to the precipitation of a crisis-like event. Unanticipated events that prompt a sudden increase in agent uncertainty are essentially 'information shocks' that have the potential to prompt a cascade of maladaptive agent responses that are 'baked in' through long term contracts and other inter-temporal mechanisms. Agent heterogeneity and inter-temporal interaction, in other words, inform how destabilising forces unfold in a market environment.

The article applies these interactionist principles to analyse markets at varying levels of aggregation. At the most fine-grained level, we have the pure simulations of Section 2, where, due to the model's purely theoretical nature, we have complete control over the states and interactions of the agents, essentially an entire economic reality within which we can explore every aspect of every agent. In Section 3, we are one step removed from the purely theoretical by introducing socio-economic data from a the Sydney housing market while maintaining complete control over the individual agents we populate the model with and the methods by which they interact with one another while still requiring that their behaviour reflects observable dynamics. In the section on financial markets, Section 4, we move even further away from the specifics of the agents, and we infer that some form of interactions are taking place between individual agents (equity traders) in the market and that we can measure these interaction and so infer a network from observed price behaviours. At the most coarse-grained level of Section 5, we look at national and international trade networks where all notion of individual agents are essentially lost and all we have available to us are the aggregate values of final demand and value added between market sectors and countries. In our final section, Section 6, we look at the role of projects as institutional structures and as mechanisms of economic development, where the relationship between multi-billion dollar projects and economic progress is writ large and is often specifically political in the way they embody and make visible the economic direction and ambitions of a nation.

Each of these economic layers has, to one degree or another, an underlying representation of heterogeneous economic agents. It is also well established in the economic literature that agent heterogeneity and inter-temporal exchange under uncertainty can propagate economic shocks through market networks [176,177]. The intellectual fine-print here is that these issues have historically been studied as autonomous problems, more often than not relegated to the realm of pure theory (although the two references provided here are notable exceptions). In cases where agent heterogeneity is considered within an inter-temporal exchange framework, analysis is typically limited by the methodological constraints of analytical tractability that is often expected in orthodox economics. In the economic discourse around crises, model tractability is a uniquely pernicious concern, as the success of an economic model has historically been evaluated in terms of the existence of a unique stable fixed-point equilibrium [49], or now more commonly, a stable steady state. Although it is uncertain how the institution of mainstream economics will shift in a post-COVID world, much of the methodological conventions considered standard economic practice developed with the belief that economies can be persistently efficient if the appropriate market mechanisms are in place. There is reason for optimism though. In a 2021 survey of articles on the economics of COVID-19 by Padhan and Prabheesh [178], they report a large number of studies using conventional economic methods (difference in difference, GARCH, descriptive statistics, etc.), but they also found methods similar to those described in this article have also been used, such as Granger causality, correlation-based minimum spanning trees, and trade network analysis (using artificial neural networks). This indicates that along with the crisis is coming a greater diversity of approaches to modelling and empirical analysis, which we would argue bodes well for both traditional economics and complexity economics.

In our opinion, CE represents an epistemological maturation of economics, in that it connects the social sciences to a broader corpus of scientific knowledge. This imposes a sufficiently theoretically agnostic and externally accepted standard of analysis to which the practice of economics can be measured against and verified by. As many of the methodological conventions within CE have been developed, applied, and verified in multiple non-economic disciplines, CE is also methodologically consistent. The principles associated with CE methods are mutually supportive and more importantly, do not typically contradict. As a result, the complexity framework is more data-orientated, tends to be testable, and is also flexible relative to, for example, the axiomatic structure of general equilibrium models. We hope and even expect that out of the current crisis will come a broader acceptance of new economic methods and theories that will have the opportunity to be developed and refined before the next crisis so that, when it does inevitably arrive, we will be better prepared with sound policy advice.

Economic Research on a Global Scale

Another very important task that has been carried out during this pandemic is the curation of data and research into central repositories for the benefit of other researchers. One such repository of economic data relevant to COVID-19 is the *Data Resources for Socioeconomic Research on COVID-19* page maintained by the European University Institute [179]. On this website can be found sections such as *Macro-financial systemic impacts* and links to key databases such as the Eurostat database that gathers statistics on the economy related to COVID-19 [180].

In addition to curating data groups, they have been curating research papers related to the economics of the pandemic, such as the work of CEPR (Centre for Economic Policy Research) that has gathered, vetted, and published COVID-19 economic papers since March 2020 [181]. The variety of subject matter and methodologies covered by the different research programs across the globe is evident in this extensive library of material and we briefly discuss four of them here. At the level of an individual's interaction with the disease and policy, vaccine policy features prominently. For example, the paper by Turner et al. [182] studies the race between the emergence of new COVID variants and the roll-out of vaccines, estimating that

> ... fully vaccinating 50% of the population would have a larger effect than simultaneously applying all forms of containment policies in their most extreme form (closure of workplaces, public transport and schools, restrictions on travel and gatherings and stay-at-home requirements). For a typical OECD country, relaxing existing containment policies would be expected to raise GDP by about 4–5%.

Stimulus packages also feature significantly in the database of articles, for example, Falcettoni and Nygaard reviewed the literature [183] on stimulus payments in the US in response, concluding in part that

> ... the poor and the young, especially those with children, should have received a larger [economic stimulus] check, which is an allocation that would have allowed for the same stimulus effect at half the cost of the actual allocation [as delivered by the US government].

Another key area of research highlighted in these articles is the impact of the pandemic on stock markets. On this topic, an article by Capelle-Blancard and Desroziers [184] showed a number of interesting results regarding the extended evolution of market response to the pandemic and the heterogeneity of its impact across 43 economies (dates in 2020):

1. Stock markets initially ignored the pandemic (until 21 February), before reacted [sic] strongly to the growing number of infected people (23 February to 20 March), while volatility surged and concerns about the pandemic arose; following the intervention of central banks (23 March to 30 April), however, shareholders no longer seemed troubled by news of the health crisis, and prices rebound all around the world.
2. Country-specific characteristics appear to have had no influence on stock market response.
3. Investors were sensitive to the number of COVID-19 cases in neighbouring but mostly wealthy countries.
4. Credit facilities and government guarantees, lower policy interest rates, and lockdown measures mitigated the decline in domestic stock prices

A final common theme reported in the CEPR database is the work on the economic impact of lockdowns. In a paper by Caselli et al. [185], they reported that their is a dual mechanism in play over the first seven months of the pandemic, one due to state enforced lockdowns and the other through voluntary social isolation in which people acted of their own accord to help mitigate the effects of social interaction on the spread of the disease. Further, they were able to estimate the differences between policies:

> We also show that lockdowns can substantially reduce COVID-19 infections, especially if they are introduced early in a country's epidemic. Despite involving short-term economic costs, lockdowns may thus pave the way to a faster recovery by containing the spread of the virus and reducing voluntary social distancing. [They were also able to show that the effect] ... entail[s] decreasing marginal economic costs but increasing marginal benefits in reducing infections.

Results such as these have a clear interaction between government economic policy and public health, a rich interaction that helps clarify many of the difficult public policy debates that often pit economic and health factors against one another.

As a final point, we consider the potential for long term economic impacts of extended health issues that are the result of so-called 'long COVID' [186]. There has been a growing awareness of the long-term negative health outcomes caused by a cluster of medical conditions such as shortness of breath, muscle aches and pains, and overall tiredness. These chronic physical manifestations of contracting COVID-19 may result in long-term reduction in individual financial stability due to the potential for job loss, long-term disability, and the increased burden of medical costs. This reduction in economic health alongside the reduction in overall quality of life may be one of the greatest long-term economic and social pitfalls of chronic COVID-19-related illnesses. It is also important to note another, less well-studied, impact of long COVID, the long-lasting cognitive deficits that come from even relatively mild symptoms of the disease. In a recent study by Hampshire et al. [187], in which 81,337 UK residents carried out a cognitive test and then reported on their COVID-19 status (asymptomatic and not biologically tested, suspected infected but not biologically tested, infected and confirmed with biological testing, admitted to hospital but not ventilated, admitted to hospital and ventilated), they found that nearly 25% of people who had contracted the disease suffered from at least some form of long COVID.

While this is concerning enough, they further showed that of those patients that were not admitted to hospital but were biologically confirmed to have had the disease the cognitive impact was the equivalent to that of having had a stroke and being admitted to hospital and ventilated was equivalent to a -7 IQ point impairment. While these are worrying results, it is slightly less worrying if the disease is confined predominantly to older members of the population where training, experience, and financial and professional stability may provide some reduction in the overall impact on quality of life and long-term financial outcomes, but it is less reassuring for younger people who are more at risk from recent variants of COVID-19. This risk led US President Joe Biden on June 18 (2021) to urge young people to be vaccinated to protect them against the new delta variant [188]. The lifetime economic and total quality of life impact of COVID-19 is not yet well understood, but for younger people the lifetime loss of earning power and productive ability is much greater than older patients, a fact that may be the cause of the most long lasting effects of this pandemic.

What we have sought to show with these case studies are some of the main effects and spillovers of a crisis and that this research is also part of a larger, diverse, and significant push to understand the global impacts of the COVID-19 pandemic. One of the points we have particularly emphasised is that, in response to crises such as this, the mainstream view of 'economic agents' needs to be broadened to fully account for the individual's biological, psychological, and sociological characteristics that underpin their fundamentally non-stationary, dynamic nature, and that, as such, economics when it is most needed during a crisis needs to account for society's distinctly non-equilibrium nature. This is almost by definition an 'out of sample' task with significant spillover effects: there is no historical context to readily draw upon and one country's changing health policy can be another country's economic crisis. Once we take these issues seriously and begin to put more significant resources to the task of understanding the complexity of our socio-economic systems there is a great deal of room for us to improve our ability to respond to future economic developments, both in and out of a crisis.

Author Contributions: Conceptualization, M.S.H.; Data curation, A.E., K.G., L.P., and S.W.; Formal analysis, M.S.H., A.E., K.G., L.P., and S.W.; Funding acquisition, M.S.H.; Investigation, M.S.H., A.E., K.G., M.H., L.P., S.W., and L.C.; Methodology, M.S.H., K.G., and S.W.; Project administration, M.S.H.; Software, K.G., L.P., and S.W.; Supervision, M.S.H.; Validation, K.G., L.P., and S.W.; Visualization, M.S.H., A.E., K.G., L.P., and S.W.; Writing–original draft, M.S.H., A.E., K.G., M.H., L.P., S.W., and L.C.; Writing–review and editing, M.S.H., and L.C. All authors have read and agreed to the published version of the manuscript.

Funding: This research received funding from the Australian Research Council grant number: DP170102927.

Data Availability Statement: All publicly available data used in this work is available at the links provided in the manuscript. Proprietary data has also been sourced from SIRCA/CoreLogic for the Australian housing ABM and used with their permission.

Acknowledgments: We would like to thank Kimberley Yoo for helping in reviewing the economic content of the article. Data on housing transactions are supplied by Securities Industry Research Centre of Asia-Pacific (SIRCA) on behalf of CoreLogic, Inc. (Sydney, Australia).

Conflicts of Interest: The authors declare no conflict of interest.

Notes

1. https://atlas.cid.harvard.edu, accessed on 13 October 2021.
2. https://oec.world/en/resources/about, accessed on 13 October 2021.
3. Also see *Australian households and businesses amass \$200 billion in savings during COVID-19 pandemic* 9 News, 14 January 2021, and *COVID-19 hit many Australians hard, but there were winners in the pandemic economy*, ABC, 23 February 2021.
4. "Australia's house prices soar to record highs over 2020", https://www.domain.com.au/news/australias-house-prices-soar-to-record-highs-over-2020-1020487/, accessed on 13 October 2021.

[5] A method for simplifying networks by using the minimum number of maximally weighted edges needed to connect all nodes without forming loops.

[6] https://www.oecd.org/sti/ind/tiva/TiVA2018_Indicators_Guide.pdf, accessed on 13 October 2021.

[7] https://www.oecd.org/sti/ind/measuring-trade-in-value-added.htm, accessed on 13 October 2021.

[8] China's and Mexico's sub-classifications are aggregated (i.e., CNH, CN1, CN2, etc.).

[9] See their website: https://covid.econ.cam.ac.uk, accessed on 13 October 2021.

[10] https://www.reuters.com/world/us/us-house-approves-715-bln-infrastructure-bill-2021-07-01/, accessed on 13 October 2021.

References

1. Arthur, W.B. Complexity economics. In *Complexity and the Economy*; Oxford University Press: Oxford, UK, 2013.
2. Hausmann, R.; Hidalgo, C.A.; Bustos, S.; Coscia, M.; Simoes, A. *The Atlas of Economic Complexity: Mapping Paths to Prosperity*; MIT Press: Cambridge, MA, USA, 2014.
3. Hidalgo, C.A.; Klinger, B.; Barabási, A.L.; Hausmann, R. The product space conditions the development of nations. *Science* **2007**, *317*, 482–487. [CrossRef]
4. Hidalgo, C.A.; Hausmann, R. The building blocks of economic complexity. *Proc. Natl. Acad. Sci. USA* **2009**, *106*, 10570–10575. [CrossRef]
5. Turrell, A. Agent-based models: Understanding the economy from the bottom up. *Bank Engl. Q. Bull.* **2016**, *4*, 173–188.
6. MacGee, J.J.; Pugh, T.M.; See, K. *The Heterogeneous Effects of COVID-19 on Canadian Household Consumption, Debt and Savings*; Technical Report; Bank of Canada: Ottawa, ON, Canada, 2020.
7. Poledna, S.; Miess, M.G.; Hommes, C.H. Economic Forecasting with an Agent-Based Model. Available online: https://ssrn.com/abstract=3484768 (accessed on 4 February 2020). [CrossRef]
8. Heise, A. Whither economic complexity? A new heterodox economic paradigm or just another variation within the mainstream? *Int. J. Plur. Econ. Educ.* **2017**, *8*, 115–129.
9. Zhou, W.X.; Sornette, D. Self-organizing Ising model of financial markets. *Eur. Phys. J. B* **2007**, *55*, 175–181. [CrossRef]
10. Schinckus, C. Ising model, econophysics and analogies. *Phys. A Stat. Mech. Its Appl.* **2018**, *508*, 95–103. [CrossRef]
11. Plerou, V.; Gopikrishnan, P.; Stanley, H.E. Two-phase behaviour of financial markets. *Nature* **2003**, *421*, 130–130. [CrossRef] [PubMed]
12. Brock, W.A.; Durlauf, S.N. Discrete choice with social interactions. *Rev. Econ. Stud.* **2001**, *68*, 235–260. [CrossRef]
13. Aoki, M. *New Approaches to Macroeconomic Modeling: Evolutionary Stochastic Dynamics, Multiple Equilibria, and Externalities as Field Effects*; Cambridge University Press: Cambridge, UK, 1998.
14. Mantegna, R.N.; Stanley, H.E. Turbulence and financial markets. *Nature* **1996**, *383*, 587–588. [CrossRef]
15. Ghashghaie, S.; Breymann, W.; Peinke, J.; Talkner, P.; Dodge, Y. Turbulent cascades in foreign exchange markets. *Nature* **1996**, *381*, 767–770. [CrossRef]
16. Milchtaich, I. Network topology and the efficiency of equilibrium. *Games Econ. Behav.* **2006**, *57*, 321–346. [CrossRef]
17. Witt, U. *Evolutionary Economics*; Edward Algar Publishing: Aldershot, UK, 1993.
18. Martin, R.; Sunley, P. Complexity thinking and evolutionary economic geography. *J. Econ. Geogr.* **2007**, *7*, 573–601. [CrossRef]
19. Scatà, M.; Di Stefano, A.; La Corte, A.; Liò, P.; Catania, E.; Guardo, E.; Pagano, S. Combining evolutionary game theory and network theory to analyze human cooperation patterns. *Chaos Solitons Fractals* **2016**, *91*, 17–24. [CrossRef]
20. Kasthurirathna, D.; Harre, M.; Piraveenan, M. Optimising influence in social networks using bounded rationality models. *Soc. Netw. Anal. Min.* **2016**, *6*, 1–14. [CrossRef]
21. Arthur, W.B. Foundations of complexity economics. *Nat. Rev. Phys.* **2021**, *3*, 136–145. [CrossRef] [PubMed]
22. Anderle, R.V. Modern Evolutionary Economics. *Rev. Bras. Inov.* **2020**, *19*, 1–3. [CrossRef]
23. Nelson, R.R.; Dosi, G.; Helfat, C.E.; Pyka, A.; Saviotti, P.P.; Lee, K.; Winter, S.G.; Dopfer, K.; Malerba, F. *Modern Evolutionary Economics: An Overview*; Cambridge University Press: Cambridge, UK, 2018.
24. Carrère, C.; Mrázová, M.; Neary, J.P. Gravity Without Apology: The Science of Elasticities, Distance and Trade. *Econ. J.* **2020**, *130*, 880–910. [CrossRef]
25. Isard, W. Location theory and trade theory: Short-run analysis. *Q. J. Econ.* **1954**, *68*, 305–320. [CrossRef]
26. Anderson, J.E. A theoretical foundation for the gravity equation. *Am. Econ. Rev.* **1979**, *69*, 106–116.
27. Bergstrand, J.H. The gravity equation in international trade: Some microeconomic foundations and empirical evidence. *Rev. Econ. Stat.* **1985**, *67*, 474–481. [CrossRef]
28. Rauch, J.E. Business and social networks in international trade. *J. Econ. Lit.* **2001**, *39*, 1177–1203. [CrossRef]
29. Rauch, J.E. Networks versus markets in international trade. *J. Int. Econ.* **1999**, *48*, 7–35. [CrossRef]
30. Chaney, T. The network structure of international trade. *Am. Econ. Rev.* **2014**, *104*, 3600–3634. [CrossRef]
31. Dalin, C.; Konar, M.; Hanasaki, N.; Rinaldo, A.; Rodriguez-Iturbe, I. Evolution of the global virtual water trade network. *Proc. Natl. Acad. Sci. USA* **2012**, *109*, 5989–5994. [CrossRef]
32. Akerman, A.; Seim, A.L. The global arms trade network 1950–2007. *J. Comp. Econ.* **2014**, *42*, 535–551. [CrossRef]
33. Edwards, J. *Reconstruction: Australia after COVID*; Penguin Random House Australia: Melbourne, Australia, 2021.

34. Smillie, D. Regional Trade Agreements. The World Bank, Brief. 2018. Available online: https://www.worldbank.org/en/topic/regional-integration/brief/regional-trade-agreements (accessed on 14 October 2021).
35. Sreedevi, R.; Saranga, H. Uncertainty and supply chain risk: The moderating role of supply chain flexibility in risk mitigation. *Int. J. Prod. Econ.* **2017**, *193*, 332–342. [CrossRef]
36. Huang, Y.; Lin, C.; Liu, S.; Tang, H. Trade Networks and Firm Value: Evidence from the US-China Trade War. CEPR Discussion Paper No. DP14173. 2019. Available online: https://ssrn.com/abstract=3504602 (accessed on 14 October 2021).
37. Levermann, A. Climate economics: Make supply chains climate-smart. *Nat. News* **2014**, *506*, 27. [CrossRef] [PubMed]
38. Watts, D.J.; Strogatz, S.H. Collective dynamics of 'small-world' networks. *Nature* **1998**, *393*, 440–442. [CrossRef] [PubMed]
39. Barabási, A.L.; Albert, R. Emergence of scaling in random networks. *Science* **1999**, *286*, 509–512. [CrossRef]
40. Bhattacharya, K.; Mukherjee, G.; Saramäki, J.; Kaski, K.; Manna, S.S. The international trade network: Weighted network analysis and modelling. *J. Stat. Mech. Theory Exp.* **2008**, *2008*, P02002. [CrossRef]
41. Barigozzi, M.; Fagiolo, G.; Mangioni, G. Identifying the community structure of the international-trade multi-network. *Phys. A Stat. Mech. Its Appl.* **2011**, *390*, 2051–2066. [CrossRef]
42. Fronczak, A.; Fronczak, P. Statistical mechanics of the international trade network. *Phys. Rev. E* **2012**, *85*, 056113. [CrossRef] [PubMed]
43. Gallegati, M.; Keen, S.; Lux, T.; Ormerod, P. Worrying trends in econophysics. *Phys. A Stat. Mech. Its Appl.* **2006**, *370*, 1–6. [CrossRef]
44. McCauley, J.L. Response to "worrying trends in econophysics". *Phys. A Stat. Mech. Its Appl.* **2006**, *371*, 601–609. [CrossRef]
45. Di Matteo, T.; Aste, T. No Worries: Trends in Econophysics. *Eur. Phys. J. B* **2007**, *55*, 121-122. [CrossRef]
46. Arthur, W.B. Complexity and the economy. *Science* **1999**, *284*, 107–109. [CrossRef]
47. LeBaron, B.; Arthur, W.B.; Palmer, R. Time series properties of an artificial stock market. *J. Econ. Dyn. Control* **1999**, *23*, 1487–1516. [CrossRef]
48. Arthur, W.B. Inductive reasoning and bounded rationality. *Am. Econ. Rev.* **1994**, *84*, 406–411.
49. Farmer, J.D.; Geanakoplos, J. The virtues and vices of equilibrium and the future of financial economics. *Complexity* **2009**, *14*, 11–38. [CrossRef]
50. Rosser, J.B., Jr. The rise and fall of catastrophe theory applications in economics: Was the baby thrown out with the bathwater? *J. Econ. Dyn. Control* **2007**, *31*, 3255–3280. [CrossRef]
51. Harré, M.; Bossomaier, T. Equity trees and graphs via information theory. *Eur. Phys. J. B* **2010**, *73*, 59–68. [CrossRef]
52. Bossomaier, T.; Barnett, L.; Harré, M. Information and phase transitions in socio-economic systems. *Complex Adapt. Syst. Model.* **2013**, *1*, 1–25. [CrossRef]
53. Kasthurirathna, D.; Piraveenan, M.; Harré, M. Influence of topology in the evolution of coordination in complex networks under information diffusion constraints. *Eur. Phys. J. B* **2014**, *87*, 1–15. [CrossRef]
54. Harré, M.S.; Prokopenko, M. The social brain: Scale-invariant layering of Erdős–Rényi networks in small-scale human societies. *J. R. Soc. Interface* **2016**, *13*, 20160044. [CrossRef]
55. Harré, M.; Bossomaier, T. Phase-transition–like behaviour of information measures in financial markets. *EPL Europhys. Lett.* **2009**, *87*, 18009. [CrossRef]
56. Wolpert, D.H.; Harré, M.; Olbrich, E.; Bertschinger, N.; Jost, J. Hysteresis effects of changing the parameters of noncooperative games. *Phys. Rev. E* **2012**, *85*, 036102. [CrossRef] [PubMed]
57. Harré, M.S.; Atkinson, S.R.; Hossain, L. Simple nonlinear systems and navigating catastrophes. *Eur. Phys. J. B* **2013**, *86*, 1–8. [CrossRef]
58. Harré, M.S.; Bossomaier, T. Strategic islands in economic games: Isolating economies from better outcomes. *Entropy* **2014**, *16*, 5102–5121. [CrossRef]
59. Harré, M.S.; Harris, A.; McCallum, S. Singularities and Catastrophes in Economics: Historical Perspectives and Future Directions. *arXiv* **2019**, arXiv:1907.05582.
60. Hopmere, M.; Crawford, L.; Harré, M.S. Proactively Monitoring Large Project Portfolios. *Proj. Manag. J.* **2020**, *51*, 656–669. [CrossRef]
61. Slavko, B.; Glavatskiy, K.; Prokopenko, M. City structure shapes directional resettlement flows in Australia. *Sci. Rep.* **2020**, *10*, 1–11. [CrossRef] [PubMed]
62. Glavatskiy, K.S.; Prokopenko, M.; Carro, A.; Ormerod, P.; Harré, M. Explaining herding and volatility in the cyclical price dynamics of urban housing markets using a large-scale agent-based model. *SN Bus. Econ.* **2021**, *1*, 1–21. [CrossRef]
63. Evans, B.P.; Glavatskiy, K.; Harré, M.S.; Prokopenko, M. The impact of social influence in Australian real estate: Market forecasting with a spatial agent-based modell. *J. Econ. Interact. Coord.* **2021**. [CrossRef]
64. Crosato, E.; Prokopenko, M.; Harré, M.S. The polycentric dynamics of Melbourne and Sydney: Suburb attractiveness divides a city at the home ownership level. *Proc. R. Soc. A* **2021**, *477*, 20200514. [CrossRef]
65. Wolpert, D.; Jamison, J.; Newth, D.; Harré, M. Strategic choice of preferences: The persona model. *BE J. Theor. Econ.* **2011**, *11*, doi:10.2202/1935-1704.1593. [CrossRef]
66. Harré, M.; Bossomaier, T.; Snyder, A. The perceptual cues that reshape expert reasoning. *Sci. Rep.* **2012**, *2*, 1–9. [CrossRef]
67. Harré, M. The neural circuitry of expertise: Perceptual learning and social cognition. *Front. Hum. Neurosci.* **2013**, *7*, 852. [CrossRef]

68. Gobet, F.; Snyder, A.; Bossomaier, T.; Harré, M. Designing a "better" brain: Insights from experts and savants. *Front. Psychol.* **2014**, *5*, 470. [CrossRef] [PubMed]
69. Prokopenko, M.; Harré, M.; Lizier, J.; Boschetti, F.; Peppas, P.; Kauffman, S. Self-referential basis of undecidable dynamics: From the Liar paradox and the halting problem to the edge of chaos. *Phys. Life Rev.* **2019**, *31*, 134–156. [CrossRef]
70. Harré, M. Entropy and Transfer Entropy: The Dow Jones and the Build Up to the 1997 Asian Crisis. In *Proceedings of the International Conference on Social Modeling and Simulation, Plus Econophysics Colloquium 2014*; Springer: Cham, Switzerland, 2015; pp. 15–25.
71. Bossomaier, T.; Barnett, L.; Harré, M.; Lizier, J.T. *An Introduction to Transfer Entropy*; Springer Int. Publ.: Cham, Switzerland, 2016; Volume 65.
72. Prokopenko, M.; Barnett, L.; Harré, M.; Lizier, J.T.; Obst, O.; Wang, X.R. Fisher transfer entropy: Quantifying the gain in transient sensitivity. *Proc. R. Soc. A Math. Phys. Eng. Sci.* **2015**, *471*, 20150610. [CrossRef]
73. Harré, M. Utility, revealed preferences theory, and strategic ambiguity in iterated games. *Entropy* **2017**, *19*, 201. [CrossRef]
74. Harré, M.S. Strategic information processing from behavioural data in iterated games. *Entropy* **2018**, *20*, 27. [CrossRef] [PubMed]
75. Bossomaier, T.; Barnett, L.; Steen, A.; Harré, M.; d'Alessandro, S.; Duncan, R. Information flow around stock market collapse. *Account. Financ.* **2018**, *58*, 45–58. [CrossRef]
76. Harré, M.S. Information Theory for Agents in Artificial Intelligence, Psychology, and Economics. *Entropy* **2021**, *23*, 310. [CrossRef] [PubMed]
77. Evans, B.; Prokopenko, M. A Maximum Entropy Model of Bounded Rational Decision-Making with Prior Beliefs and Market Feedback. *Entropy* **2021**, *23*, 669. [CrossRef] [PubMed]
78. Gallegati, M.; Palestrini, A.; Rosser, J.B. The period of financial distress in speculative markets: Interacting heterogeneous agents and financial constraints. *Macroecon. Dyn.* **2011**, *15*, 60. [CrossRef]
79. Rosser, J.B.; Rosser, M.V.; Gallegati, M. A Minsky-Kindleberger perspective on the financial crisis. *J. Econ. Issues* **2012**, *46*, 449–458. [CrossRef]
80. Gordon, M.J. Towards a theory of financial distress. *J. Financ.* **1971**, *26*, 347–356. [CrossRef]
81. Corbae, D.; Quintin, E. Leverage and the foreclosure crisis. *J. Political Econ.* **2015**, *123*, 1–65. [CrossRef]
82. Haughwout, A.; Lee, D.; Tracy, J.S.; Van der Klaauw, W. Real Estate Investors, the Leverage Cycle, and the Housing Market Crisis. FRB of New York Staff Report, No. 514. 2011. Available online: https://ssrn.com/abstract=1926858 (accessed on 14 October 2021). [CrossRef]
83. Bookstaber, R.M.; Paddrik, M.E. An Agent-Based Model for Crisis Liquidity Dynamics. Office of Financial Research Working Paper, No. 15-18. 2015. Available online: https://ssrn.com/abstract=2664230 (accessed on 14 October 2021). [CrossRef]
84. Bookstaber, R.; Paddrik, M.; Tivnan, B. An agent-based model for financial vulnerability. *J. Econ. Interact. Coord.* **2018**, *13*, 433–466. [CrossRef]
85. Kindleberger, C.P.; Aliber, R.Z. *Manias, Panics and Crashes: A History of Financial Crises*; Palgrave Macmillan: London, UK, 2011.
86. Australian Government Department of Health. *Coronavirus (COVID-19) Advice for International Travellers*; Australian Government Department of Health: Canberra, Australia, 2021.
87. Pacific, C.A. *CoreLogic Quarterly Rental Review*; CoreLogic: Sydney, Australia, 2020.
88. Reserve Bank of Australia. *Term Funding Facility–Reduction in Interest Rate to Further Support the Australian Economy*; Media Releases–RBA: Sydney, Australia, 2020.
89. Pacific, C.A. *Housing Market Update*; CoreLogic: Sydney, Australia, 2020.
90. Australian Taxation Office A.G. JobKeeper Payment. 2021. Available online: https://www.ato.gov.au/general/jobkeeper-payment/ (accessed on 14 October 2021).
91. Services Australia A.G. JobSeeker Payment. 2021. Available online: https://www.servicesaustralia.gov.au/individuals/services/centrelink/jobseeker-payment (accessed on 14 October 2021).
92. Jribi, S.; Ismail, H.B.; Doggui, D.; Debbabi, H. COVID-19 virus outbreak lockdown: What impacts on household food wastage? *Environ. Dev. Sustain.* **2020**, *22*, 3939–3955. [CrossRef] [PubMed]
93. Australian Bureau of Statistics. Household Impacts of COVID-19 Survey. 2021. Available online: https://www.abs.gov.au/statistics/people/people-and-communities/household-impacts-covid-19-survey/apr-2021 (accessed on 14 October 2021).
94. Australian Bureau of Statistics. Insights of the Housing Market During COVID-19. 2021. Available online: https://www.abs.gov.au/articles/insights-housing-market-during-covid-19 (accessed on 14 October 2021).
95. Lux, T.; Marchesi, M. Volatility clustering in financial markets: A microsimulation of interacting agents. *Int. J. Theor. Appl. Financ.* **2000**, *3*, 675–702. [CrossRef]
96. Gopikrishnan, P.; Plerou, V.; Amaral, L.A.N.; Meyer, M.; Stanley, H.E. Scaling of the distribution of fluctuations of financial market indices. *Phys. Rev. E* **1999**, *60*, 5305. [CrossRef]
97. Lux, T. Turbulence in financial markets: The surprising explanatory power of simple cascade models. *Quant. Financ.* **2001**, *1*, 632–640. [CrossRef]
98. Zhou, W.X. Finite-size effect and the components of multifractality in financial volatility. *Chaos Solitons Fractals* **2012**, *45*, 147–155. [CrossRef]
99. Gidea, M.; Katz, Y. Topological data analysis of financial time series: Landscapes of crashes. *Phys. A Stat. Mech. Its Appl.* **2018**, *491*, 820–834. [CrossRef]

100. Jiang, Z.Q.; Xie, W.J.; Zhou, W.X.; Sornette, D. Multifractal analysis of financial markets: A review. *Rep. Prog. Phys.* **2019**, *82*, 125901. [CrossRef]
101. Padash, A.; Chechkin, A.V.; Dybiec, B.; Pavlyukevich, I.; Shokri, B.; Metzler, R. First-passage properties of asymmetric Lévy flights. *J. Phys. A Math. Theor.* **2019**, *52*, 454004. [CrossRef]
102. Arias-Calluari, K.; Alonso-Marroquin, F.; Harré, M.S. Closed-form solutions for the Lévy-stable distribution. *Phys. Rev. E* **2018**, *98*, 012103. [CrossRef] [PubMed]
103. Alonso-Marroquin, F.; Arias-Calluari, K.; Harré, M.; Najafi, M.N.; Herrmann, H.J. Q-Gaussian diffusion in stock markets. *Phys. Rev. E* **2019**, *99*, 062313. [CrossRef] [PubMed]
104. Arias-Calluari, K.; Alonso-Marroquin, F.; Najafi, M.N.; Harré, M. Methods for forecasting the effect of exogenous risks on stock markets. *Phys. A Stat. Mech. Its Appl.* **2021**, *568*, 125587. [CrossRef]
105. Arias-Calluari, K.; Najafi, M.N.; Harré, M.; Yang, Y.; Alonso-Marroquin, F. Testing stationarity of the detrended price return in stock markets. *Phys. A Stat. Mech. Its Appl.* **2021**, 126487, doi:10.1016/j.physa.2021.126487 [CrossRef]
106. Plerou, V.; Gopikrishnan, P.; Rosenow, B.; Amaral, L.A.N.; Guhr, T.; Stanley, H.E. Random matrix approach to cross correlations in financial data. *Phys. Rev. E* **2002**, *65*, 066126. [CrossRef]
107. Plerou, V.; Gopikrishnan, P.; Rosenow, B.; Amaral, L.A.N.; Stanley, H.E. Universal and nonuniversal properties of cross correlations in financial time series. *Phys. Rev. Lett.* **1999**, *83*, 1471. [CrossRef]
108. Onnela, J.P.; Chakraborti, A.; Kaski, K.; Kertesz, J. Dynamic asset trees and Black Monday. *Phys. A Stat. Mech. Its Appl.* **2003**, *324*, 247–252. [CrossRef]
109. Onnela, J.P.; Chakraborti, A.; Kaski, K.; Kertesz, J.; Kanto, A. Asset trees and asset graphs in financial markets. *Phys. Scr.* **2003**, *2003*, 48. [CrossRef]
110. Onnela, J.P.; Chakraborti, A.; Kaski, K.; Kertesz, J.; Kanto, A. Dynamics of market correlations: Taxonomy and portfolio analysis. *Phys. Rev. E* **2003**, *68*, 056110. [CrossRef]
111. Onnela, J.P.; Saramäki, J.; Kertész, J.; Kaski, K. Intensity and coherence of motifs in weighted complex networks. *Phys. Rev. E* **2005**, *71*, 065103. [CrossRef]
112. Barnett, L.; Lizier, J.T.; Harré, M.; Seth, A.K.; Bossomaier, T. Information flow in a kinetic Ising model peaks in the disordered phase. *Phys. Rev. Lett.* **2013**, *111*, 177203. [CrossRef]
113. Granger, C.W. Testing for causality: A personal viewpoint. *J. Econ. Dyn. Control* **1980**, *2*, 329–352. [CrossRef]
114. Schreiber, T. Measuring information transfer. *Phys. Rev. Lett.* **2000**, *85*, 461. [CrossRef]
115. Simoes, A.J.G.; Hidalgo, C.A. The economic complexity observatory: An analytical tool for understanding the dynamics of economic development. In Proceedings of the Workshops at the Twenty-Fifth AAAI Conference on Artificial Intelligence, Palo Alto, CA, USA, 7–11 August 2011.
116. Hartmann, D.; Guevara, M.R.; Jara-Figueroa, C.; Aristarán, M.; Hidalgo, C.A. Linking economic complexity, institutions, and income inequality. *World Dev.* **2017**, *93*, 75–93. [CrossRef]
117. Lapatinas, A.; Garas, A.; Boleti, E.; Kyriakou, A. Economic Complexity and Environmental Performance: Evidence from a World Sample. *Environ. Mod. Ass.* **2021**, *26*, 251–270.
118. Adam, A.; Garas, A.; Katsaiti, M.S.; Lapatinas, A. Economic complexity and jobs: An empirical analysis. *Econ. Innov. New Technol.* **2021**, 1–28. [CrossRef]
119. Dolfing, A.G.; Leuven, J.R.; Dermody, B.J. The effects of network topology, climate variability and shocks on the evolution and resilience of a food trade network. *PLoS ONE* **2019**, *14*, e0213378. [CrossRef]
120. Fair, K.R.; Bauch, C.T.; Anand, M. Dynamics of the global wheat trade network and resilience to shocks. *Sci. Rep.* **2017**, *7*, 1–14. [CrossRef]
121. Schwartz, E.; Smith, J.E. Short-term variations and long-term dynamics in commodity prices. *Manag. Sci.* **2000**, *46*, 893–911. [CrossRef]
122. Devalkar, S.K.; Anupindi, R.; Sinha, A. Integrated optimization of procurement, processing, and trade of commodities. *Oper. Res.* **2011**, *59*, 1369–1381. [CrossRef]
123. Demirer, M.; Diebold, F.X.; Liu, L.; Yilmaz, K. Estimating global bank network connectedness. *J. Appl. Econom.* **2018**, *33*, 1–15. [CrossRef]
124. Gemünden, H.G.; Lehner, P.; Kock, A. The project-oriented organization and its contribution to innovation. *Int. J. Proj. Manag.* **2018**, *36*, 147–160. [CrossRef]
125. Flyvbjerg, B. What you Should Know about Megaprojects and Why: An Overview. *Proj. Manag. J.* **2014**, *45*, 6–19. [CrossRef]
126. Ansar, A.; Flyvbjerg, B.; Budzier, A.; Lunn, D. Should we build more large dams? The actual costs of hydropower megaproject development. *Energy Policy* **2014**, *69*, 43–56. [CrossRef]
127. Flyvbjerg, B.; Stewart, A. *Olympic Proportions: Cost and Cost Overrun at the Olympics 1960–2012*; SSRN Scholarly Paper ID 2238053; Social Science Research Network: Rochester, NY, USA, 2012. [CrossRef]
128. Olds, K. Globalization and the production of new urban spaces: Pacific Rim megaprojects in the late 20th century. *Environ. Plan. A* **1995**, *27*, 1713–1743. [CrossRef]
129. Zeković, S.; Maričić, T.; Vujošević, M. Megaprojects as an instrument of urban planning and development: Example of Belgrade Waterfront. In *Technologies for Development: From Innovation to Social Impact*; Springer: Cham, Switzerland, 2018; pp. 153–164. doi: 10.1007/978-3-319-91068-0. [CrossRef]

130. Söderlund, J.; Sankaran, S.; Biesenthal, C. *The Past and Present of Megaprojects*; SAGE Publications: Los Angeles, CA, USA, 2017. [CrossRef]
131. Devenow, A.; Welch, I. Rational herding in financial economics. *Eur. Econ. Rev.* **1996**, *40*, 603–615. [CrossRef]
132. Parker, W.D.; Prechter, R.R. Herding: An Interdisciplinary Integrative Review from a Socionomic Perspective. 2005. Available online: https://ssrn.com/abstract=2009898 (accessed on 14 October 2021). [CrossRef]
133. Scheffer, M.; Bascompte, J.; Brock, W.A.; Brovkin, V.; Carpenter, S.R.; Dakos, V.; Held, H.; Van Nes, E.H.; Rietkerk, M.; Sugihara, G. Early-warning signals for critical transitions. *Nature* **2009**, *461*, 53–59. [CrossRef]
134. Scheffer, M. Foreseeing tipping points. *Nature* **2010**, *467*, 411–412. [CrossRef]
135. Seekell, D.A.; Carpenter, S.R.; Pace, M.L. Conditional heteroscedasticity as a leading indicator of ecological regime shifts. *Am. Nat.* **2011**, *178*, 442–451. [CrossRef]
136. Dai, L.; Vorselen, D.; Korolev, K.S.; Gore, J. Generic indicators for loss of resilience before a tipping point leading to population collapse. *Science* **2012**, *336*, 1175–1177. [CrossRef]
137. Slavko, B.; Glavatskiy, K.S.; Prokopenko, M. Revealing configurational attractors in the evolution of modern Australian and US cities. *Chaos Solitons Fractals* **2021**, *148*, 111079. [CrossRef]
138. Staniek, M.; Lehnertz, K. Symbolic transfer entropy. *Phys. Rev. Lett.* **2008**, *100*, 158101. [CrossRef]
139. Yao, W.; Wang, J. Multi-scale symbolic transfer entropy analysis of EEG. *Phys. A Stat. Mech. Its Appl.* **2017**, *484*, 276–281. [CrossRef]
140. Wang, L.; Xue, W.; Li, Y.; Luo, M.; Huang, J.; Cui, W.; Huang, C. Automatic epileptic seizure detection in EEG signals using multi-domain feature extraction and nonlinear analysis. *Entropy* **2017**, *19*, 222. [CrossRef]
141. Lenton, T.; Livina, V.; Dakos, V.; Van Nes, E.; Scheffer, M. Early warning of climate tipping points from critical slowing down: comparing methods to improve robustness. *Philos. Trans. R. Soc. A Math. Phys. Eng. Sci.* **2012**, *370*, 1185–1204. [CrossRef] [PubMed]
142. Tylianakis, J.M.; Coux, C. Tipping points in ecological networks. *Trends Plant Sci.* **2014**, *19*, 281–283. [CrossRef] [PubMed]
143. Dakos, V.; Matthews, B.; Hendry, A.P.; Levine, J.; Loeuille, N.; Norberg, J.; Nosil, P.; Scheffer, M.; De Meester, L. Ecosystem tipping points in an evolving world. *Nat. Ecol. Evol.* **2019**, *3*, 355–362. [CrossRef] [PubMed]
144. Barnett, L.; Barrett, A.B.; Seth, A.K. Granger causality and transfer entropy are equivalent for Gaussian variables. *Phys. Rev. Lett.* **2009**, *103*, 238701. [CrossRef] [PubMed]
145. Lizier, J.T. JIDT: An information-theoretic toolkit for studying the dynamics of complex systems. *Front. Robot. AI* **2014**, *1*, 11. [CrossRef]
146. Fagiolo, G.; Reyes, J.; Schiavo, S. On the topological properties of the world trade web: A weighted network analysis. *Phys. A Stat. Mech. Its Appl.* **2008**, *387*, 3868–3873. [CrossRef]
147. Maeng, S.E.; Choi, H.W.; Lee, J.W. Complex networks and minimal spanning trees in international trade network. In *International Journal of Modern Physics: Conference Series*; World Scientific: Singapore, 2012; Volume 16, pp. 51–60.
148. Barzel, B.; Barabási, A.L. Universality in network dynamics. *Nat. Phys.* **2013**, *9*, 673–681. [CrossRef]
149. Ariu, A. Crisis-proof services: Why trade in services did not suffer during the 2008–2009 collapse. *J. Int. Econ.* **2016**, *98*, 138–149. [CrossRef]
150. del Rio-Chanona, R.M.; Mealy, P.; Pichler, A.; Lafond, F.; Farmer, J.D. Supply and demand shocks in the COVID-19 pandemic: An industry and occupation perspective. *Oxf. Rev. Econ. Policy* **2020**, *36*, S94–S137. [CrossRef]
151. Kelly, G.; La Cava, G. Value-added Trade and the Australian Economy. *RBA Bull.* **2013**. Available online: https://www.rba.gov.au/publications/bulletin/2013/mar/ (accessed on 14 October 2021).
152. Johnson, R.C.; Noguera, G. Accounting for intermediates: Production sharing and trade in value added. *J. Int. Econ.* **2012**, *86*, 224–236. [CrossRef]
153. Deguchi, T.; Takahashi, K.; Takayasu, H.; Takayasu, M. Hubs and authorities in the world trade network using a weighted HITS algorithm. *PLoS ONE* **2014**, *9*, e100338.
154. Zhu, Z.; Cerina, F.; Chessa, A.; Caldarelli, G.; Riccaboni, M. The rise of China in the international trade network: A community core detection approach. *PLoS ONE* **2014**, *9*, e105496. [CrossRef]
155. Escaith, H.; Lindenberg, N.; Miroudot, S. *International Supply Chains and Trade Elasticity in Times of Global Crisis*; World Trade Organization (Economic Research and Statistics Division) Staff Working Paper ERSD-2010-08; WTO: Geneva, Switzerland, 2010.
156. Turner, J. *The Handbook of Project-Based Management*, 2nd ed.; McGraw-Hill: London, UK, 1999.
157. Geraldi, J.; Maylor, H.; Williams, T.M. Now, let's make it really complex (complicated): A systematic review of the complexities of projects. *Int. J. Oper. Prod. Manag.* **2011**, *31*, 966–990. [CrossRef]
158. Chipulu, M.; Neoh, J.G.; Ojiako, U.; Williams, T. A Multidimensional Analysis of Project Manager Competences. *IEEE Trans. Eng. Manag.* **2013**, *60*, 506–517. [CrossRef]
159. Project Management Institute. *PMI Lexicon of Project Management Terms*; Technical Report; Project Management Institute: Newtown Square, PA, USA, 2017.
160. Spanuth, T.; Heidenreich, S.; Wald, A. Temporary organisations in the creation of dynamic capabilities: Effects of temporariness on innovative capacity and strategic flexibility. *Ind. Innov.* **2020**, *27*, 1186–1208. [CrossRef]
161. Schoper, Y.G.; Wald, A.; Ingason, H.T.; Fridgeirsson, T.V. Projectification in Western economies: A comparative study of Germany, Norway and Iceland. *Int. J. Proj. Manag.* **2018**, *36*, 71–82. [CrossRef]

162. Turner, J.; Keegan, A. Mechanisms of governance in the project-based organization:: Roles of the broker and steward. *Eur. Manag. J.* **2001**, *19*, 254–267. [CrossRef]
163. Caldas, C.; Gupta, A. Critical factors impacting the performance of mega-projects. *Eng. Constr. Archit. Manag.* **2017**, *24*, 920–934. [CrossRef]
164. Flyvbjerg, B.; Bruzelius, N.; Rothengatter, W. *Megaprojects and Risk: An Anatomy of Ambition*; Cambridge University Press: New York, NY, USA, 2003.
165. Bodrožić, Z.; Adler, P.S. The Evolution of Management Models: A Neo-Schumpeterian Theory. *Adm. Sci. Q.* **2018**, *63*, 85–129. [CrossRef]
166. Hanusch, H.; Pyka, A. Principles of Neo-Schumpeterian Economics. *Camb. J. Econ.* **2006**, *31*, 275–289. [CrossRef]
167. Locatelli, G. Why are Megaprojects, Including Nuclear Power Plants, Delivered Overbudget and Late? Reasons and Remedies. *arXiv* **2018**, arXiv:1802.07312.
168. Aritua, B.; Smith, N.J.; Bower, D. Construction client multi-projects—A complex adaptive systems perspective. *Int. J. Proj. Manag.* **2009**, *27*, 72–79. [CrossRef]
169. Bousquin, J. $9.6B Worth of Infrastructure Projects Delayed or Canceled during COVID-19, 2020. Published 4 August 2020. Available online: https://www.constructiondive.com/news/96b-worth-of-infrastructure-projects-delayed-or-canceled-during-covid-19/582713/ (accessed on 14 October 2021).
170. Premo Black, A. *Impacts of COVID-19 on State and Local Transportation Revenues & Construction Programs*; Technical Report; American Road & Transportation Builders Association, 2020. Available online: https://www.artba.org/wp-content/uploads/20 20/11/202010.23_ARTBA_COVID19RevImpact_v22.pdf (accessed on 14 October 2021).
171. Hopkirk, E. Projects Cancelled as Workloads and Income Hit by COVID-19 Pandemic, RIBA Survey Finds. 2020. Available online: https://www.building.co.uk/news/projects-cancelled-as-workloads-and-income-hit-by-covid-19-pandemic-riba-survey-finds/5105430.article (accessed on 14 October 2021).
172. Steeman, M. Almost 200 Construction Projects Impacted by COVID-19, New Report Says | Stuff.co.nz, 2020. 9 April 2020. Available online: https://www.stuff.co.nz/business/property/120894743/almost-200-construction-projects-impacted-by-covid1 9-new-report-says (accessed on 14 October 2021).
173. ConstructConnect. *Delayed Projects Report-9 July 2021*; Technical Report; Construct Connect: Cincinnati OH, USA, 2021. Available online: https://www.constructconnect.com/contact-us (accessed on 14 October 2021).
174. AGC. *State List of Construction Employment by Metropolitan Area or Division, September 2019–September 2020*; Technical Report; AGC The Construction Association: Arlington, VA, USA, 2020.
175. Jones, K. *State of the Construction Industry: One Year into the Pandemic*; Construct Connect: Cincinnati OH, USA, 2021. Available online: https://www.constructconnect.com/blog/state-of-the-construction-industry-one-year-into-the-pandemic (accessed on 14 October 2021).
176. Christelis, D.; Georgarakos, D.; Jappelli, T. Wealth shocks, unemployment shocks and consumption in the wake of the Great Recession. *J. Monet. Econ.* **2015**, *72*, 21–41. [CrossRef]
177. Lupu, I. From the Financial Crisis to the Real Economy: The main channels of transmission through a theoretical perspective. *Euro Econ.* **2012**, *31*, 45–50.
178. Padhan, R.; Prabheesh, K. The economics of COVID-19 pandemic: A survey. *Econ. Anal. Policy* **2021**, *70*, 220–237. [CrossRef]
179. European University Institute | COVID-19 Research. 2021. Available online: https://www.eui.eu/Research/Library/ResearchGuides/Economics/Crisis (accessed on 14 October 2021).
180. Eurostat | COVID-19 | Economy. 2021. Available online: https://ec.europa.eu/eurostat/web/covid-19/economy (accessed on 14 October 2021).
181. CEPR | Covid Economics. 2021. Available online: https://cepr.org/content/covid-economics-vetted-and-real-time-papers-0 (accessed on 14 October 2021).
182. Turner, D.; Égert, B.; Guillemette, Y.; Botev, J. *The Tortoise and the Hare: The Race between Vaccine Rollout and New COVID Variants*; OECD Economics Department Working Papers. 2021. Available online: https://www.oecd-ilibrary.org/economics/the-tortoise-and-the-hare-the-race-between-vaccine-rollout-and-new-covid-variants_4098409d-en (accessed on 14 October 2021).
183. Falcettoni, E.; Nygaard, V.M. A literature review on the impact of increased unemployment insurance benefits and stimulus checks in the United States. *Covid Econ.* **2021**, *64*, 186–201.
184. Capelle-Blancard, G.; Desroziers, A. The stock market is not the economy? Insights from the COVID-19 crisis. In *Insights from the COVID-19 Crisis (June 16, 2020). CEPR COVID Economics*; VoxEU & CEPR Coverage of the Covid-19 Global Pandemic. 2020. Available online: https://voxeu.org/article/stock-market-and-economy-insights-covid-19-crisis (accessed on 14 October 2021).
185. Caselli, F.; Grigoli, F.; Sandri, D. Protecting lives and livelihoods with early and tight lockdowns. *BE J. Macroecon.* **2021**. [CrossRef]
186. Mahase, E. COVID-19: What do we know about "long covid"? *BMJ* **2020**, *370*.
187. Hampshire, A.; Trender, W.; Chamberlain, S.R.; Jolly, A.E.; Grant, J.E.; Patrick, F.; Mazibuko, N.; Williams, S.C.; Barnby, J.M.; Hellyer, P.; et al. Cognitive deficits in people who have recovered from COVID-19. *EClinicalMedicine* **2021**, *39*, 101044.
188. Kupferschmidt, K.; Wadman, M. Delta Variant Triggers New Phase in the Pandemic. *Science* **2021**, *372*, 6549. [CrossRef]

MDPI
St. Alban-Anlage 66
4052 Basel
Switzerland
Tel. +41 61 683 77 34
Fax +41 61 302 89 18
www.mdpi.com

Systems Editorial Office
E-mail: systems@mdpi.com
www.mdpi.com/journal/systems

www.ingramcontent.com/pod-product-compliance
Lightning Source LLC
LaVergne TN
LVHW070907170726
843515LV00005B/723